WHITE DWARFS: COSMOLOGICAL AND GALACTIC PROBES

AF213606

ASTROPHYSICS AND SPACE SCIENCE LIBRARY

VOLUME 332

EDITORIAL BOARD

Chairman

W.B. BURTON, National Radio Astronomy Observatory, Charlottesville, Virginia, U.S.A.
(bburton@nrao.edu); University of Leiden, The Netherlands (burton@strw.leidenuniv.nl)

Executive Committee

J. M. E. KUIJPERS, *University of Nijmegen, The Netherlands*
E. P. J. VAN DEN HEUVEL, *University of Amsterdam, The Netherlands*
H. VAN DER LAAN, *University of Utrecht, The Netherlands*

MEMBERS

J. N. BAHCALL, *The Institute for Advanced Study, Princeton, U.S.A.*
F. BERTOLA, *University of Padua, Italy*
J. P. CASSINELLI, *University of Wisconsin, Madison, U.S.A.*
C. J. CESARSKY, *European Southern Observatory, Garching bei München, Germany*
O. ENGVOLD, *University of Oslo, Norway*
A. HECK, *Strassbourg Astronomical Observatory, France*
R. McCRAY, *University of Colorado, Boulder, U.S.A.*
P. G. MURDIN, *Institute of Astronomy, Cambridge, U.K.*
F. PACINI, *Istituto Astronomia Arcetri, Firenze, Italy*
V. RADHAKRISHNAN, *Raman Research Institute, Bangalore, India*
K. SATO, *School of Science, The University of Tokyo, Japan*
F. H. SHU, *University of California, Berkeley, U.S.A.*
B. V. SOMOV, *Astronomical Institute, Moscow State University, Russia*
R. A. SUNYAEV, *Space Research Institute, Moscow, Russia*
Y. TANAKA, *Institute of Space & Astronautical Science, Kanagawa, Japan*
S. TREMAINE, *Princeton University, U.S.A.*
N. O. WEISS, *University of Cambridge, U.K.*

WHITE DWARFS: COSMOLOGICAL AND GALACTIC PROBES

Edited by

EDWARD M. SION

Villanova University, Villanova, U.S.A.

STÉPHANE VENNES

Johns Hopkins University, Baltimore, MD, U.S.A.

and

HARRY L. SHIPMAN

University of Delaware, Newark, DE, U.S.A.

Springer

A C.I.P. Catalogue record for this book is available from the Library of Congress.

ISBN-13 978-90-481-6932-0 (PB)
ISBN-13 978-1-4020-3725-2 (e-book)

Published by Springer,
P.O. Box 17, 3300 AA Dordrecht, The Netherlands.

www.springeronline.com

Printed on acid-free paper

Cover picture: White Dwarfs in Globular Cluster Mu,
Hubble Space Telescope, NASA and H. Richer (University of British Columbia)

All Rights Reserved
© 2010 Springer
No part of this work may be reproduced, stored in a retrieval system, or transmitted
in any form or by any means, electronic, mechanical, photocopying, microfilming, recording
or otherwise, without written permission from the Publisher, with the exception
of any material supplied specifically for the purpose of being entered
and executed on a computer system, for exclusive use by the purchaser of the work.

Printed in the Netherlands.

Contents

Preface

In the past few years, general astronomical interest has concentrated on several objects and phenomena where white dwarf stars play a key role. Type Ia supernovae have been used as evidence to show that, in fact, Einstein did not make his greatest blunder when he allowed for the possibility of a cosmological constant. Improvements in our knowledge of the Hubble parameter have revived interest in the use of white dwarf stars as a different type of cosmochronometer to measure the age of the Galaxy and thus set constraints on the age of the Universe. In roughly the same time period, there have been considerable advances in our understanding of white dwarf stars, both as isolated stars in the field and as members of interacting binary systems. Much of this advance has come from the availability of spacecraft observations from missions like *HST*, *IUE*, *ROSAT*, *EXOSAT*, *Chandra*, *ORFEUS*, *EUVE*, *HUT*, and *FUSE*. The discovery of thousands of new white dwarfs from a number of large surveys and the potential of the Sloan Digital Sky Survey have added impetus to the field.

Studies of interacting binaries such as classical novae, supersoft X-ray binaries, symbiotic variables, dwarf novae and nova-like objects have revealed the differences between the thermal evolution of single and close binary systems as well as heightened interest in these systems as progenitors of the cosmologically important Type I supernovae. A more speculative question is whether classical novae can be understood well enough that they might provide another kind of standard candle. The role of competing physical processes (mixing versus diffusion) in determining chemical abundances at the onset of, and in the ejecta of, the thermonuclear explosion must be understood not only for classical novae but also for accreting X-ray bursting neutron stars in X-ray binaries.

This book, and the conference that gave rise to it, focus on aspects of white dwarfs that are providing new information as a result of the large ground-based surveys and new space results, and how this information can be used to gain understanding of the age of the galactic disk, Type Ia supernovae, and the physics and phenomenology of the cataclysmic variable stars which, according to current conventional wisdom, a wide range of phenomena from accretion effects in nearby close binaries to local open clusters to distant globular clusters and dark matter in the halo and ultimately to set a lower limit to the age of our galaxy and the universe.

This book: Going Beyond the Conference

This book includes some extensive review articles, in some cases by invitation to scientists who were not present at the conference itself. At the conference, one of us (Shipman) presented some brief review talks based on the published literature. We also did not limit our authors to a fixed number of pages, thus encouraging authors to present longer, more discursive review material. As a result, we believe that people who read this book will get a stronger and broader perspective than those who simply went to the conference. Readers who did not participate in the conference can gain this broad perspective on several issues that are critically important for astronomy today.

The common thread binding this book is the role of white dwarf research in solving problems in cosmology, galactic structure, and the physics of accretion. The book naturally divides into three major sections: I. Halo White Dwarfs and Galactic Structure; II. Type Ia Supernovae; and III. Cataclysmic Variables and White Dwarf Accretion. The overlap in these sections simply reflects the impetus behind the conference and the book. That is, to bring out the ways in which current research on white dwarf stars is impacting on problems in cosmology, galactic structure, late stellar evolution and accretion onto white dwarfs in interacting binaries. To help guide the reader, we have designated several articles which offer a broader perspective, either solicited by invitation of the editors or contributed, as being "review papers". To this end, we point out the articles by G. Fontaine et al. in section I., the articles by A.V. Filippenko, and C.A. Tout in Section II. and by Warner and Woudt, P. and Szkody et al. in section III.

Giving credit where credit is due: The four authors of this Introduction all played different roles in this enterprise as they functioned as a team. The original idea of proposing such a conference came from Paula Szkody and Ed Sion. Harry Shipman joined up and became principal proposal author and impresario for the conference, with assistance from Sion and Vennes. Sion initiated the possiblity of a volume beyond the abbreviated form of an IAU transaction and became the principal editor, with assistance from Vennes and Shipman. The four of us gratefully acknowledge financial support for this work from NASA and from the National Science Foundation. H.L. Shipman and E.M. Sion acknowledge support from the Delaware Space Grant Program, which has facilitated their collaboration.

H.L. SHIPMAN. E.M. SION, P. SZKODY, AND S. VENNES

SECTION I. HALO WHITE DWARFS AND GALACTIC STRUCTURE

OLD ULTRACOOL WHITE DWARFS AS COSMOLOGICAL PROBES

Review paper

G. Fontaine, P. Bergeron, P. Brassard
Département de Physique, Université de Montréal
fontaine@astro.umontreal.ca, bergeron@astro.umontreal.ca, brassard@astro.umontreal.ca

Abstract We briefly review the status of the considerable efforts that have been made in recent years to detect and characterize populations of white dwarfs that would be sufficiently old to be of cosmological interest. Except for the well documented turnover in the luminosity function of local white dwarfs at low temperatures which is directly related to the finite age of the galactic disk, the optimistic view that many shared several years ago has subdued somewhat as these populations have remained elusive. We argue that fast moving white dwarfs in the local neighborhood uncovered during the course of several recent proper motion surveys most likely do not belong to an old population and, therefore, do not bear the signature of the age of the halo. This would still leave the MACHO results, if interpreted as true halo events, as indicative of the possibility of an old white dwarf population lurking in the halo of our galaxy. Attempts to detect directly this putative population through proper motion studies in the Hubble Deep Field have so far failed. We point out that completely invisible white dwarfs, ultracool "red" He-atmosphere halo stars, could also contribute to the microlensing events. Finally, we discuss the impressive efforts made recently to unveil the old ultracool white dwarfs that must certainly exist in the globular cluster M4. On the basis of our atmosphere and cooling models, we find that such a population has probably not been detected yet, however. This may imply that a recent estimate of the age of the cluster based on white dwarf cosmochronology is premature.

Keywords: Cosmochronology, Galactic halo, Galactic disk

1. Introduction

In the last several years there has been considerable research devoted to the uncovering of old ultracool ($T_{\rm eff} \lesssim$ 4000 K) white dwarfs for the purposes of assessing their potential contribution to baryonic dark matter and of using them as age indicators of various old stellar systems. Their complete absence in our local region of space has led to a firm estimate of the age of the disk

E.M. Sion, S. Vennes and H.L. Shipman (eds.), White Dwarfs: Cosmological and Galactic Probes, 3-13.
© 2005 *Springer.*

(see, e.g., Leggett, Ruiz, & Bergeron 1998 and references therein). The tantalizing possibility that a significantly older population of white dwarfs lurks in the halo has also been raised on the basis of various interpretations of a number of observational discoveries. Thus, several contemporaneous proper motion surveys have revealed the existence of high-velocity white dwarfs in our neighborhood, which have been interpreted as interlopers from the halo (e.g., Oppenheimer et al. 2001 and references therein). Furthermore, Hansen (1998) suggested that very old halo white dwarfs could possibly be associated with a group of unidentified blue objects in the Hubble Deep Field (HDF). This suggestion gave support to an interpretation of the MACHO results in terms of old white dwarfs being the lensing objects in the halo of our galaxy (Alcock et al. 2000). Finally, the prospects of detecting the very old ultracool white dwarfs that must certainly exist in globular clusters and using them as age indicators appeared within reach following the exploratory work of Richer et al. (1997) on M4 using the Hubble Space Telescope (HST).

These exciting results have led to a renewed interest in white dwarf cooling calculations and model atmosphere calculations using upgraded input physics and extending into the regime of very cool, evolved white dwarfs. A complete review of these developements has been presented recently by Fontaine, Brassard, & Bergeron (2001). In that paper (and see also Fontaine 2001), a rather optimistic assessment of the prospects for white dwarf cosmochronology was given. We review critically here where we now stand on that front a few years later.

2. Halo Candidates and White Dwarfs in the Solar Neighborhood

Liebert, Dahn, & Monet (1989) were the first to identify white dwarfs in our local region of space that could be interpreted as interlopers from the halo on the basis of their kinematic properties. Since then, several other proper motion surveys have revealed many other candidate halo stars (Hambly et al. 1999; Ibata et al. 2000; De Jong, Kuijken, & Neeser 2000; Oppenheimer et al. 2001). The Oppenheimer et al. (2001) survey, in particular, has been the most ambitious of those and has uncovered some 38 objects which have been interpreted as cool halo white dwarfs. That work has generated a flurry of activity which is still being pursued (see, e.g., Salim et al. 2004).

Bergeron (2003) has recently presented a critical examination of these halo white dwarf candidates and concluded that most of the fast movers are too warm and most likely too young to belong to the galactic halo. Instead, those objects appear to be relatively young disk stars with high velocities with respect to the local standard of rest. Possible mechanisms to produce fast moving white dwarfs in our region of space have been proposed by Davies, King, & Ritter

(2002), Hansen (2003), and Koopmans & Blandford (2004). We build in the rest of this section on the work of Bergeron (2003).

We first concentrate on those objects for which reliable parallax measurements are available as well as complete energy distributions extending into the infrared (see Bergeron 2003). And indeed, the most direct way for estimating the ages of individual white dwarfs is to compare theoretical isochrones with observational data points in a mass-effective temperature diagram. For cool and very cool stars, the spectroscopic lines have all but disappeared and the mass can be estimated only if a parallax is known. This implies that the method is restricted to nearby objects.

Figure 1 illustrates theoretical isochrones in a $M - T_{\rm eff}$ diagram for three different sets of models. One of our purposes here is to remind the reader that white dwarf cosmochronology is based on assumptions about the core composition and the chemical layering of the envelope of the models used (see Fontaine et al. 2001 for further details). In that connection, the models of Fontaine et al. (2001) have pure C cores and, as such, evolve the slowest due to their increased heat capacity. It was repeatedly stated in that paper that the age estimates obtained there are simply illustrative examples and correspond to *upper limits* for the true ages. We use more realistic C/O cores here and the lower panel of Figure 1, corresponding to "thick" H envelope models, should be compared to Figure 8 of Fontaine et al. (2001) to assess the (important) effects of changing the core composition. For its part, the upper panel in comparison to the lower one illustrates the dramatic effect of getting rid of the insulating layer of hydrogen in white dwarfs. And indeed, because of the extreme transparency of pure He at low temperatures, a cool white dwarf of a given mass reaches a given effective temperature much earlier if it has a pure He envelope as opposed to a stratified envelope with atop a layer of hydrogen. The middle panel illustrates an intermediate situation for layered models with "thin" H envelopes.

Figure 1 further shows that isochrones are sensitive functions of the mass of a white dwarf, a basic fact that seems to be forgotten sometimes in the literature. Their "S" shape is due, at the top, to the effects of crystallization and Debye cooling which manifest themselves at higher values of $T_{\rm eff}$ in the more massive stars. At the bottom, the finite lifetimes of the progenitors on the main sequence – longer for the less massive stars – enter into the picture. Hence a cool or ultracool white dwarf does not necessarily have to be old if it is massive enough. Conversely, a warm white dwarf does not have to be young but, in that case, its mass has to fall into a very narrow range, between $\sim$0.45 and perhaps $\sim$0.50 $M_\odot$ for the particular choice of the initial-to-final mass relationship that we used in our models. This is why age determinations of individual cool white dwarfs must rely on reliable estimates of both the effective temperature *and* the mass.

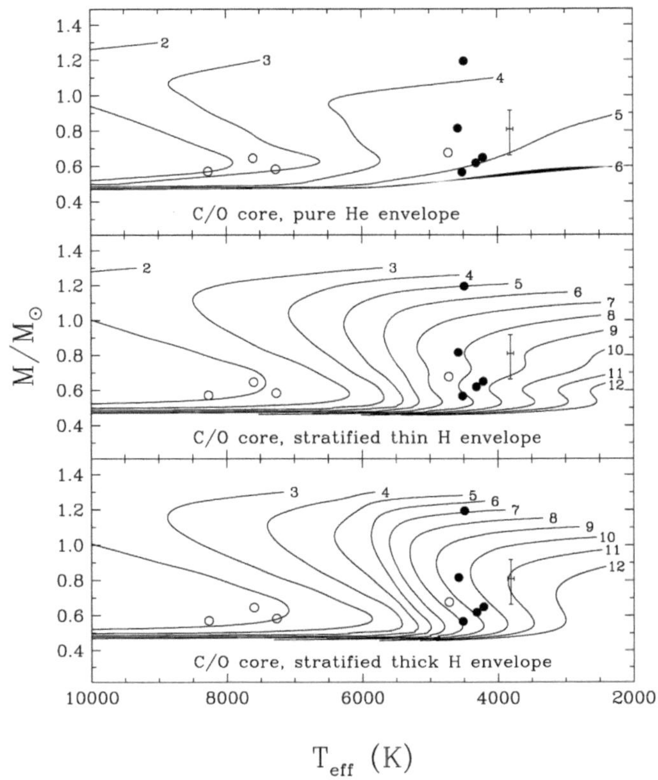

Figure 1. White dwarf isochrones in the $M - T_{\rm eff}$ diagram. Each isochrone is expressed in units of Gyr. Each panel refers to evolutionary models with the same C/O cores but with different envelope layering. The lower panel represents models with a "thick" H envelope, i.e., log $(M(\mathrm{H})/\mathrm{M}_*) \equiv$ log q(H) = -4 and log q(He) = -2. The middle panel is for "thin" H envelope models, i.e., with log q(H) = -10 (and log q(He) = -2). The upper panel refers to pure He envelope models (with log q(He) = -2). The figure illustrates the considerable influence that the compositional layering has on the ages of white dwarfs. The various data points shown in the diagram are discussed in the text.

The best characterization of the cool white dwarf population in our neighborhood has been carried out by Bergeron, Ruiz, & Leggett (1997) and Bergeron, Leggett, & Ruiz (2001). The stars in their samples all belong to the local disk according to their kinematic properties, with the exception of the handful of high-velocity stars found in the Liebert et al. (1989) study and also analyzed by Bergeron et al. (1997, 2001). We isolated the five coolest ($T_{\rm eff}$ < 4600 K) single degenerates in their parallax sample and plotted them in our Figure 1 (the solid circles). We note that no ultracool ($T_{\rm eff} < 4000$ K) white dwarf has been found by Bergeron et al. and that the simplest interpretation is that the disk is too young to have produced many of those cooled off objects. Quite certainly, if a nonnegligible fraction of white dwarfs evolve as pure He envelope models, ultracool specimens must have been produced during the life of the disk, but those objects evolve so quickly that their space density must be relatively low between $T_{\rm eff}$ ~4000 K and the current detection limit (~2000 K?) and, therefore, they must be difficult to find in large numbers.

If we assume that the white dwarfs in the Bergeron et al. sample have evolved mostly as thick H envelope models (with C/O cores), then the lower panel of Figure 1 suggests an age for the disk of about 9.5 Gyr, consistent with the estimate of Bergeron et al. (2001) based on the same models as those considered here. This uses the two oldest stars in the sample according to our isochrones and should be compared to the estimate of ~11 Gyr derived earlier by Fontaine et al. (2001) under the same conditions but on the basis of pure C core models believed to be less realistic than the C/O models used here. Note that the cool but rather massive star shown in the panel (it is ESO 439−26) is a relatively young object with an estimated age of about 6.8 Gyr despite having an extremely low luminosity of log $L/L_\odot = -4.94$. On the other hand, if the bulk of the white dwarfs in the disk have evolved as thin H envelope models (middle panel), then the age of the disk would rather be closer to 8.5 Gyr, again in line with the earlier estimate of Bergeron et al. (2001) on the basis of these same other models and perfectly consistent with the determination of Leggett et al. (1998) based on a match with the observational luminosity function.

We also indicated the locations (the open circles) of four high proper motion white dwarfs culled from the original sample of Liebert et al. (1989) and analyzed by Bergeron et al. (2001) who combined complete energy distributions and recent parallax measurements to derive masses and effective temperatures. It is clear that those objects, even though they are fast movers, cannot belong to an old halo population; they are neither old nor ultracool. Of course, if the halo "turnoff" mass was associated to a white dwarf mass significantly larger than the cutoff mass of 0.45 $M_\odot$ used in our models, then the remnants shown in the figure (open circles) could be much older than indicated here. However, there is no a priori reason to believe that. A better possibility is that provided by WD 0346+246 (the cross in Figure 1), a high velocity local white dwarf discovered by Hambly et al. (1999; see also Hodgkin et al. 2000) in their own proper motion survey. This star was reanalyzed by Bergeron (2001) who found a mixed atmosphere with N(He)/N(H) = 1.3, $T_{\rm eff}$ = 3780 K, and log g = 8.34. While WD 0346+246 is indeed an ultracool white dwarf by our above criterion, its age is unfortunately very uncertain because of its mixed chemical composition. If we make abstraction of that fact for a moment and assume that it has cooled as a thick H envelope white dwarf, then the lower panel of Figure 1 would suggest indeed that WD 0346+246 is significantly older than the disk. Combined with its halo kinematic characteristics, this would make that star an ideal representative of the elusive population of old halo white dwarfs. Unfortunately, its mixed atmospheric composition implies at the very least that WD 0346+246 is not currently cooling as a H-atmosphere white dwarf, and that it probably did not in the past. Hence, it is most likely much younger than the age indicated by its position in the lower panel of the figure. The fact of the matter is that the perfect halo candidate, a H-atmosphere white dwarf that would be both ultracool and show high spatial velocity has yet to be found (see Bergeron & Leggett 2002 for further discussion on this).

We mentioned at the beginning of the section that the most extensive proper motion survey has been that of Oppenheimer et al. (2001) which led to the discovery of 38 high velocity white dwarfs in the local neighborhood. Unfortunately, the vast majority of these objects have no parallax measurements nor measured energy distributions so comparisons with isochrones such as those provided in Figure 1 cannot be made. While progress on that front is being made (see, e.g., Salim et al. 2004), Bergeron (2003) provided a critical analysis of these obervations and excluded the possibility of an old population for most of the stars in the Oppenheimer et al. sample. Again, provided the hypothetical burst of stellar formation in the young halo did not lead to unusually massive white dwarf remnants (via an exceptionally shallow initial-to-final mass relationship for example), we conclude that there is as yet no firm evidence for the presence of old halo white dwarfs zooming by us in our local region of space (but see Méndez 2002).

3. Ultracool White Dwarfs in Distant Systems?

The first results of the MACHO microlensing experiment provided indirect evidence for the existence of an old population of white dwarfs in the galactic halo. This generated intense interest in the field and many attempts were made to characterize such a population (see, e.g., Chabrier 1999 or Saumon & Jacobson 1999 and references therein). After many years of observations, the MACHO team has ruled out a dark halo made entirely of dark baryonic matter, but still favors a halo comprising up to 20% of MACHOs, each with a typical mass of 0.5 $M_{\odot}$ (Alcock et al. 2000). An upper limit of 30% for such stellar-mass halo objects has been derived in EROS, another ongoing microlensing experiment (Lasserre et al. 2000). The most natural candidates for the MACHOs, subluminous objects with masses of half the solar mass, remain of course ultracool white dwarfs that would be the remnants of an initial burst of star formation when the halo was created.

Quite a bit of enthusiasm was raised when Hansen (1998; and see also Chabrier, Segretain, & Méra 1996) suggested that this "dark"halo matter could be seen directly in the form of very old H-atmosphere white dwarfs making up some of the so-called blue unidentified objects in the HDF. A similar suggestion came from Méndez & Minniti (2000) working with the HDF South. The expectations were further raised when Ibata et al. (1999), using second epoch exposures, reported the probable discovery of detectable proper motions in up to five of these "blue unidentified objects" in the HDF as would be expected of halo stars residing at kiloparsec distances. Fontaine et al. (2001) analyzed the limited available photometry of these extremely faint objects ($I \sim 28$) and concluded that two of them, 4−492 and 4−551, have energy distributions compatible with those of very cool H-atmosphere white dwarfs while the other three do not. This was perfectly in line with the expectations of Chabrier et al. (1996) and Hansen (1998), the latter estimating that two or three cool halo white dwarfs should be found in the narrow HDF. Unfortunately, third epoch HDF exposures (Richer 2003) did not confirm the proper motions initially reported by Ibata et al. (1999), so that the evidence for the "direct" detection of ultracool halo white dwarfs has all but evaporated (but see Nelson et al. 2002). We are left with the puzzle provided by the MACHO events, assuming that the lensing objects indeed belong to the halo (see, e.g., Sahu 1994).

Have old ultracool white dwarfs been detected in other distant stellar systems? We expect that such old stars indeed exist in globular clusters and that their detection would provide a potentially very interesting and independent way for estimating the ages of the clusters. Following the initial work of Richer et al. (1997), M4 became the most promising candidate and substantial efforts were made to obtain deeper exposures with the HST. These efforts, based on some 123 orbits, led to the publication of Hansen et al. (2002) who claimed to have determined an age for M4 of 12.7 ± 0.7 Gyr on the basis of their observationally derived white dwarf luminosity function. Our Figure 1 clearly shows that, at this age, the vast majority of the white dwarfs present in M4 must indeed be ultracool.

As impressive as this result appears to be, it was recently reexamined by De Marchi et al. (2004) who, in our view, raised a number of valid arguments. Among those, two were previously addressed in Fontaine et al. (2001) in their discussion of the application of white dwarf cosmochronology to stellar clusters. Firstly, it was pointed out there that the main signature of the finite age of a cluster from a white dwarf point of view in a CMD is the location of a pileup of objects along the cooling sequence accompanied by a gradual decrease in star density with fainter magnitudes. In the cases where the pileup is located beyond the limiting magnitude, only lower limits to the ages can be obtained. Two examples were provided by Fontaine et al. (2001), 1) an estimate of the age of the old open cluster M67 could be obtained from the CFHT observations of Richer et al. (1998) because the pileup of cluster white dwarfs was neatly detected as shown in their Figure 12, and 2) in contrast, from their Figure 14, only a lower limit of ~9 Gyr (it would have been even less on the basis of the more realistic C/O core models used here) could be derived for the age of the globular cluster M4 since the HST observations of Richer et al. (1997) were not deep enough to reveal the pileup.

Secondly, in a related issue, Fontaine et al. (2001) also pointed out that the differential luminosity function of a white dwarf population has the same *shape* on the high luminosity side of the maximum in the distribution whatever its age and, furthermore, that the age "information" is really on the low luminosity side of that peak, again raising the important issue of having observations deep enough to reveal the pileup of white dwarfs. De Marchi et al. (2004) reanalyzed independently the archived HST data originally obtained by Hansen et al. (2002) and found both in their CMD and in their derived luminosity function that the peak amplitude was not reached, thus allowing them to only set a lower limit of ~9 Gyr to the age of M4. They concluded that observations several magnitudes deeper than the limit obtained in the HST observations of Hansen et al. (2002) would be required to determine properly the true age of M4 on the basis of white dwarf cosmochronology.

The work of De Marchi et al. (2004) was rather harshly criticized in rebuttal papers presented by Hansen et al. (2004) and Richer et al. (2004). Among other things, Richer et al. (2004) showed that, through sheer technical prowess in the data analysis phase, they were able to push the detection limit almost a full magnitude fainter than in the more conservative study of De Marchi et al. (2004). Hansen et al. (2004) also introduced a rather elaborate and sophisticated way of comparing the observational luminosity function with model predictions. Not only did they exploit the 2D dependence of the luminosity function in the CMD, but also the absolute value (as opposed to the shape) of that function to confirm the age estimate obtained earlier in Hansen et al. (2002). In view of the importance of this result, and we concur with De Marchi et al. (2004) on this point, it seems essential that independent confirmations be brought forward.

In this spirit, and to add our grain of salt to the debate, we have compared in Figure 2 some model isochrones with the M4 data of Hansen et al. (2002) expressed in the original HST colors and kindly made available to us by Harvey Richer. This has the advantage that we do not need to worry about uncertain color transformations to a standard system. In this comparison we took great care to adopt the same minimum white dwarf mass (0.55 $M_\odot$) as Hansen et al. (2004) and we also folded in our calculations their particular choice of the initial-to-final mass relation and their prescription for the main sequence lifetime estimates. Figure 2 leaves us quite puzzled, however. There is no obvious sign of the expected pileup of stars above the detection limit even

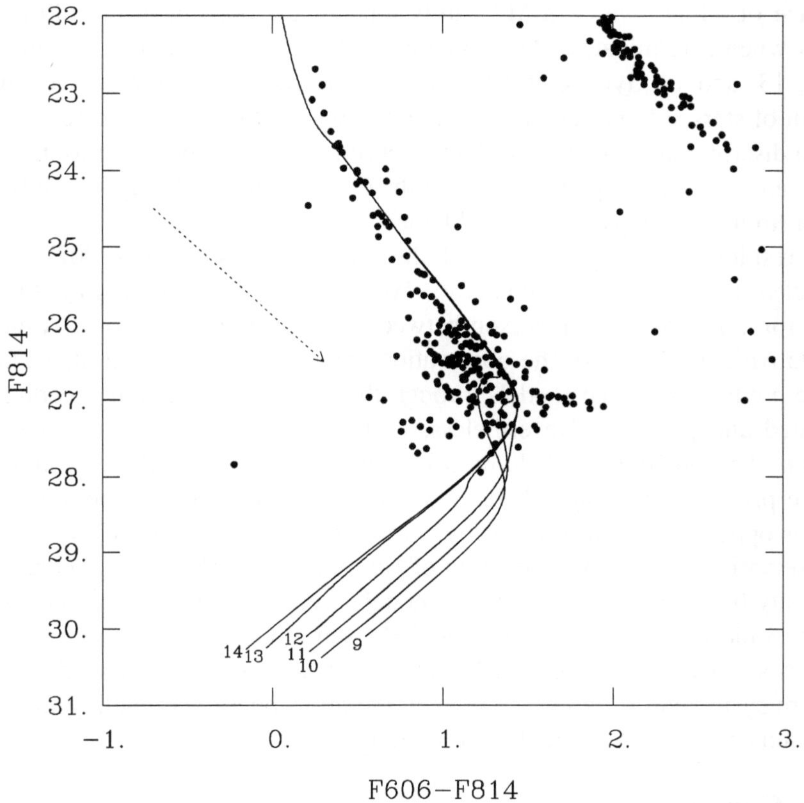

Figure 2. Color-magnitude diagram in the original HST bandpasses for the white dwarf candidates found in M4 by Hansen et al. (2002). An original distance modulus of 12.0 was used for the F814 magnitude, and a color excess of 0.41 was applied to the theoretical color index $(F606-F814)_0$. In addition, a slight adjustment was made along the reddening vector (dotted line) to make sure that the 0.55 $M_\odot$ model track would best match the data points in the $23.5 \lesssim$ F814 $\lesssim 25.0$ range. The isochrones (solid curves) are based on our C/O core white dwarf models with thick H envelopes. They are expressed in units of Gyr, and are fused together and overlap with the evolutionary track of the 0.55 $M_\odot$ model at bright magnitudes.

though it is almost a magnitude fainter than that obtained in De Marchi et al. (2004). Hansen et al. (2004) actually acknowledge that, but claim that it is not necessary to reach the pileup to derive an accurate age. What we find more worrisome however is that, according at least to our models, the observed sequence of white dwarfs in M4 barely touches the domain of old ultracool objects when it terminates. As is plainly evident in Figure 2, our isochrones for 12, 13, and 14 Gyr are still fused together when the observed sequence runs out of stars at faint magnitudes. In other words, the actual data cannot be used to discriminate between our late isochrones, so an upper limit to the age of M4 cannot be derived on the basis of the CMD shown in the figure. At best, a lower limit of perhaps 10 Gyr is obtained.

Our result is obviously at odds with the claim of Hansen et al. (2002) and it is not clear at this stage why this is so. We certainly admit that Figure 2 only allows for a qualitative comparison between theory and observations. A more quantitative comparison of the observations reported in Hansen et al. (2002) and the predictions of our models through the luminosity function is certainly warranted and planned. Nevertheless, it should be clear that isochrone fitting *must* be consistent with luminosity function studies, and we would be most surprised, if not astonished, if we were able to derive a true age estimate (as opposed to a lower limit) by comparing, even in a sophisticated way, the observational luminosity function of Hansen et al. (2002) with theoretical luminosity functions coming from the models used in Figure 2. In this connection, it would have been enlightening if Hansen et al. (2002, 2004) had also shown isochrones in their papers. For the time being, we can only suspect that there are systematic differences between the atmosphere and cooling models of Hansen and ours. We stand by our models.

4. Conclusion

Despite the considerable efforts that have been made in recent years to detect and characterize populations of white dwarfs that would be sufficiently old to be of cosmological interest, we must conclude that such populations have remained elusive so far. We argued that fast moving white dwarfs discovered recently in the local neighborhood, and interpreted by many as interlopers from the halo, most likely do not belong to an old population. Moreover, we reminded the reader that direct evidence for the existence of very old H-atmosphere ("blue") white dwarfs in the HDF North has not materialized despite earlier high hopes to the contrary. If the MACHO results are to be explained in terms of true halo events, however, then we must keep looking. In this connection, we feel that He-atmosphere white dwarfs may have been dismissed too summarily. The argument has been that all red objects in the HDF have been accounted for in terms of standard stellar populations, but a population of He-atmosphere white dwarfs resulting from an initial burst of star formation in the halo would now be so faint that it could not contribute to the red star counts in the HDF anyway. Hence, it may not be too far-fetched to suggest that truly dark He-atmosphere white dwarfs lurk in the halo and are responsible for some of the MACHO events. Finally, we argued that the population of old ultracool white dwarfs that must certainly exist in the globular cluster M4 may not have been detected yet. This may imply that a recent estimate of the age of the cluster based on white dwarf cosmochronology is premature.

References

Alcock, C. et al. 2000, *ApJ*, **541**, 281.
Bergeron, P. 2001, *ApJ*, **558**, 369.
Bergeron, P. 2003, *ApJ*, **586**, 201.
Bergeron, P., & Leggett, S.K. 2002, *ApJ*, **590**, 1070.
Bergeron, P., Leggett, S.K., & Ruiz, M.T. 2001, *ApJS*, **133**, 413.
Bergeron, P., Ruiz, M.T., & Leggett, S.K. 1997, *ApJS*, **108**, 339.
Chabrier, G. 1999, *ApJ*, **513**, 216.
Chabrier, G., Segretain, L, & Méra, D. 1996, *ApJ*, **468**, L21.
Davies, M.B., King, A., & Ritter, H. 2002, *MNRAS*, **333**, 463.
De Jong, J., Kuijken, K., & Neeser, M. 2000, /astroph/0009058.
De Marchi, G., Paresce, F., Straniero, O., & Prado Moroni, P.G. 2004, *A&A*, **415**, 971.
Fontaine, G. 2001, in *ASP Conf. Ser. 245*, Astrophysical Ages and Time Scales, ed. T. von Hippel, C. Simpson, & N. Manset (San Francisco: ASP), 173.
Fontaine, G., Brassard, P., & Bergeron, P. 2001, *PASP*, **113**, 409.
Hambly, N.C. et al. 1999, *MNRAS*, **309**, L33.
Hansen, B.M.S. 1998, *Nature*, **394**, 860.
Hansen, B.M.S. 2003, *ApJ*, **582**, 915.
Hansen, B.M.S. et al. 2002, *ApJ*, **574**, L158.
Hansen, B.M.S. et al. 2004, /astroph/0401443.
Hodgkin, S.T. et al. 2000, *Nature*, **403**, 57.
Ibata, R. et al. 1999, *ApJ*, **524**, L95.
Ibata, R. et al. 2000, *ApJ*, **532**, L41.
Koopmans, L.V.E., & Blandford, R.D. 2004, *MNRAS*, in press.
Lasserre, T. et al. 2000, *A&A*, **355**, L39.
Leggett, S.K., Ruiz, M.T., & Bergeron, P. 1998, *ApJ*, **497**, 294.
Liebert, J., Dahn, C.C., & Monet, D.G. 1989 in IAU Colloq. 114, White Dwarfs, ed. G. Wegner (Berlin: Springer), 15.
Méndez, R. 2002, *A&A*, **395**, 779.
Méndez, R., & Minniti, D. 2000, *ApJ*, **529**, 911.
Nelson, C. et al. 2002, *ApJ*, **573**, 644.
Oppenheimer, B.R., Hambly, N.C., Digby, A.P., Hodgkin, S.T., & Saumon, D. 2001, *Science*, **292**, 698.
Richer, H.B. 2003, in The Dark Universe: Matter, Energy, & Gravity, ed. M. Livio (Cambridge: Camb. U. Press), 24.
Richer, H.B. et al. 1997, *ApJ*, **484**, 741.
Richer, H.B. et al. 1998, *ApJ*, **504**, L91.
Richer, H.B. et al. 2004, /astroph/0401446.
Sahu, K.C. 1994, *Nature*, **370**, 275.
Salim S. et al. 2004, *ApJ*, **601**, 1075.
Saumon, D., & Jacobson, S.B. 1999, *ApJ*, **511**, L107.

NUMBER COUNTS OF WHITE DWARFS: THE IMPACT OF GAIA

Enrique Garcia-Berro[1], Santiago Torres[1,2], Francesca Figueras [2,3], Jordi Isern[2,4]

[1]*Departament de Física Aplicada, Universitat Politècnica de Catalunya, Escola Politécnica Superior de Castelldefels*
Av. del Canal Olímpic s/n, 08860, Castelldefels, Spain

[2]*Institut d'Estudis Espacials de Catalunya*
Edifici Nexus, Gran Capità 2-4, 08034 Barcelona, Spain

[3]*Departament de d'Astronomia i Meteorologia, Universitat de Barcelona, Facultat de Física*
Martí i Franquès 1, 08028 Barcelona, Spain

[4]*Institut de Ciències de l'Espai, CSIC*
Edifici Nexus, Gran Capità 2-4, 08034 Barcelona, Spain

Abstract In the next years, several space missions will be devoted to measure with very high accuracy the motions of a sizeable fraction of the stars of our Galaxy. The most promising one is the ESA astrometric satellite GAIA, which will provide very precise astrometry ($< 10\,\mu$as in parallax and $< 10\,\mu$as yr^{-1} in proper motion at $V \sim 15$, increasing to 0.2 mas yr^{-1} at $V \sim 20$) and multicolor photometry, for all 1.3 billion objects to $V \sim 20$, and radial velocities with accuracies of a few km s^{-1} for most stars brighter than $V \sim 17$. Consequently, full homogeneous six-dimensional phase-space information for a huge number of white dwarfs will become available. Our Monte Carlo simulator has been used to estimate the number of white dwarfs potentially observable by GAIA. Our results show which could be the impact of a mission like GAIA in the current understanding of our Galaxy. Scientific attainable goals include, among others, a reliable determination of the age of our galactic disk, a better knowledge of the structure of the halo of the Milky Way or the reconstruction of the past history of the Star Formation Rate of the galactic disk.

Keywords: Galactic disk, halo, white dwarfs

1. Introduction

GAIA is the European Space Agency (ESA) astrometric satellite now selected as a Cornerstone 6 mission as part of the Science Program (see, for instance, the URL: `http://www.rssd.esa.int/Gaia/index.html`, for a complete technical and scientific information). GAIA is an ambitious project

E.M. Sion, S. Vennes and H.L. Shipman (eds.), White Dwarfs: Cosmological and Galactic Probes, 15-24.
© 2005 *Springer.*

which will provide us with multi-color, multi-epoch photometry, astrometry and spectroscopy for all objects brighter than $V \approx 20$ (Perryman et al. 2001). The dataset generated by GAIA will be very large, providing us with accurate information of unprecedented precision of more than a billion objects in our Galaxy.

In fact, GAIA will be the successor of the astrometric satellite *Hipparcos*, which was operative from 1989 to 1993. The scientific program of *Hipparcos* was much more modest than that of GAIA since it measured the positions and proper motions of only 10^5 rather than 10^9 galactic objects. Moreover, *Hipparcos* operated on the basis of an input catalogue. Instead, GAIA will continuously scan the sky and, hence, it will determine its own targets. Finally, the scientific products of *Hipparcos* were released only when the mission was complete, whereas some of the scientific data that GAIA will collect will be partially released during the duration of the mission.

As previously mentioned, GAIA will continuously scan the sky. This will be achieved through the following procedure. GAIA rotates slowly around its spin axis, which itself precesses at a fixed angle to the Sun of $55°$. As GAIA rotates, the light of the sources is collected by three telescopes and directed towards the focal planes where the instruments are located. GAIA will have three focal planes, two of which (ASTRO-1, ASTRO-2) measure the positions of stars, whereas the third one (SPECTRO) performs spectroscopy of selected sources. Additionally GAIA will have a medium band photometer (MBP) and a broad band photometer (BBP). The telescopes are of moderate size, with no specific manufacturing complexity. The CCDs in the astrometric focal plane measure the direct stellar images in the so-called time-delayed integration (TDI) mode. GAIA observes along great circles in two simultaneous fields of view, separated by a well-known angle. The astrometric parameters of stars are derived from the analysis of the time series of the one-dimensional transits distributed over the five year mission lifetime. The orbit of GAIA will be a Lissajous-type orbit at the lagrangian point L2.

This paper assesses the number of white dwarfs potentially observable by GAIA. In doing this our Monte Carlo simulator (García–Berro et al. 1999; Torres et al. 1998) has been used. Our results show which could be the impact of a mission like GAIA in the current understanding of our Galaxy. This work is organized as follows. In §2 a brief description of our Monte Carlo simulator is given. Section 3 is devoted to analyze the results of our simulations, including the completeness of the samples of disk and halo white dwarfs, and the accuracy of the astrometric determinations of both samples. Finally, in §4 our conclusions are summarized.

2. The Monte Carlo simulator

Since our Monte Carlo simulator has been thouroughly described in previous papers (García–Berro et al. 1999; García–Berro et al. 2004) we will only summarize here the most important inputs. In our Monte Carlo simulations we have adopted a disk age of 13 Gyr. White dwarfs have been distributed according to a double exponential density law with a scale length $L = 3.5$ kpc in the galactocentric distance and with a scale height $h = 500$ pc perpendicular to the Galactic plane. A standard IMF and a constant volumetric SFR were adopted. The velocities have been derived taking into account the differential rotation of the Galaxy, the peculiar velocity of the sun and a dispersion law depending of the scale height. An accurate model of the galaxy absorption has been used as well (Hakkila et al. 1997). On the other hand, the halo was assumed to be formed 14 Gyr ago in an intense burst of star formation of duration 1 Gyr. The synthetic white dwarfs have been distributed according to a typical isothermal spherically symmetric halo. A standard IMF was adopted as well. The velocities were randomly drawn according normal distributions and adopting a rotation velocity of 250 km/s.

The procedure to obtain the synthetic stars is the following. First, we randomly choose the three-dimensional coordinates of each star of the sample according to the previously mentioned distributions. Afterwards we draw another pseudo-random number in order to obtain the main sequence mass of each star, according to the IMF. Once the mass of the progenitor of the white dwarf is known we randomly choose the time at which each star was born, according to the SFR of the population under study. Given the age of the corresponding population and the main sequence lifetime as a function of the mass in the main sequence (Iben & Laughlin 1989) we know which stars have had time enough to become white dwarfs, and given a set of cooling sequences (Salaris et al. 2000) and the initial to final mass relationship (Iben & Laughlin 1989), which are their luminosities and magnitudes. The magnitude is then converted to the instrumental magnitude of GAIA, G, which is related to the standard colors $(V, V-I)$ by the expression:

$$G = V + 0.51 - 0.50\sqrt{0.6 + (V - I - 0.6)^2} - 0.065\,(V - I - 0.6)^2$$

which is valid for $-0.4 < V - I < 6$. For $-0.4 < V - I < 1.4$, $G \approx V$. The typical errors in parallax depend on the magnitude (Perryman 2002) and are then computed using the following fit:

$$\sigma_\pi \simeq (7 + 105z + 1.3z^2 + 6\,10^{-10}z^6)^{1/2}\,[0.96 + 0.04(V - I)]$$

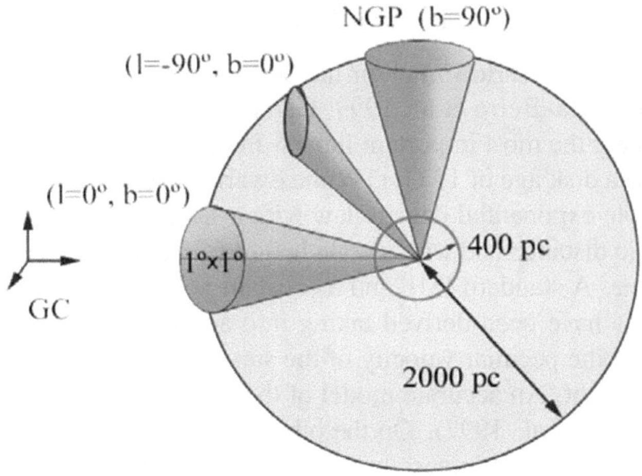

Figure 1. The adopted geometry for the Monte Carlo simulations.

Table 1. Number of disk white dwarfs for different regions of the sky, see text for details.

	$G < 20$	$G < 21$
$(l = -90°, b = 0°)$	4	9
$(l = 0°, b = 0°)$	4	8
$(b = 90°)$	11	28
All sky	$2.3 \cdot 10^5$	$6.7 \cdot 10^5$

where $z = 10^{0.4(G-15)}$. On the other hand the errors in parallax, σ_π, position σ_0, and proper motion, σ_μ, are related by: $\sigma_0 = 0.87\sigma_\pi$ and $\sigma_\mu = 0.75\sigma_\pi$, respectively.

In this way we end up with all the relevant information necessary to assess the performance of GAIA. Finally, it is important to mention at this point that since the number of synthetic stars necessary to simulate the whole Galaxy is prohibitively large we have only determined the number of white dwarfs that could be detected in a small window of $1° \times 1°$ for each of the three directions shown in Figure 1. The number density of white dwarfs for each of these pencils was normalized to the local observed density of either disk or halo white dwarfs. After doing this we average our results over the whole sky.

3. Expected number of disk and halo white dwarfs

The total number of disk white dwarfs along the three above mentioned pencils and the total number of white dwarfs accessible to GAIA are shown in Table 1 for two different limiting magnitudes.

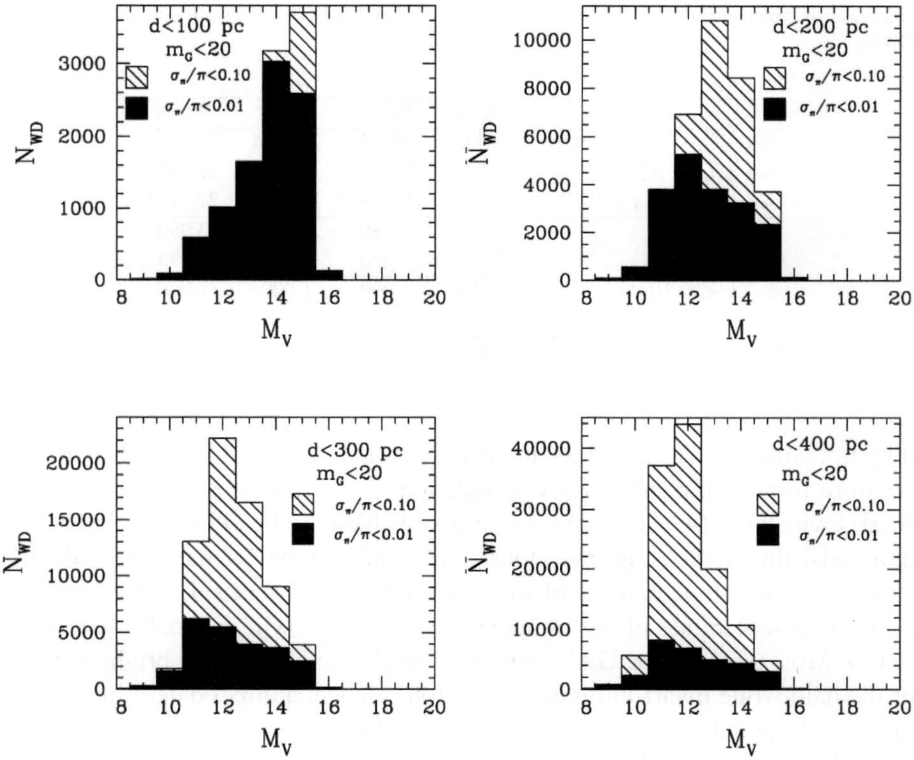

Figure 2. Distribution of the number of disk white dwarfs detected by GAIA as a function of their absolute magnitude according to their errors in parallax, for different distances.

In Figure 2 we show the distribution in the number of disk white dwarfs detected by GAIA as a function of their absolute magnitude, according to their errors in parallax, for 100, 200, 300 and 400 pc. The total estimated number of white dwarfs within those distances can be found in the first row of Table 2. Of these white dwarfs, those which pass the cut in G apparent magnitude are given in the second row of Table 2. The third row lists which of these will also have measurable proper motions. The completitude of the sample necessary to build the white dwarf luminosity function is assessed in the last row of this Table. The second and the third section of Table 2 assess the accuracy of the astrometric measurements. As it can be seen there most of the detected white dwarfs will have good determinations ($\sigma < 0.1$) for both the parallax and the proper motion up to distances of more than 400 pc and superb accuracies, $\sigma < 0.01$, for half of the sample will be obtained up to distances of about 200 pc.

The same exercise can be done for the halo white dwarf population. However the results are not as encouraging as those obtained for the disk white

Table 2. Results of the Monte Carlo simulations of disk white dwarf Population accessible to GAIA.

	100 pc	200 pc	300 pc	400 pc
$N_{\rm WD}(9.0 < G < 29.0)$	11595	80775	273684	743893
$N_{\rm WD}(G < 21)$	11593	57804	121474	226922
$N_{\rm WD}(\mu > \mu_{\rm cut})$	11593	57804	121472	226919
$\sigma_\mu/\mu < 0.10$	1.000	0.999	0.999	0.998
$\sigma_\mu/\mu < 0.01$	0.998	0.966	0.929	0.893
$\sigma_\pi/\pi < 0.10$	1.000	1.000	1.000	1.000
$\sigma_\pi/\pi < 0.01$	0.877	0.559	0.357	0.249
η	0.999	0.716	0.444	0.305

dwarf population. These results are displayed in Table 3 and Figure 3. Perhaps the most important result of these simulations is that the number of halo white dwarfs which GAIA will be able to observe is likely to be of the order of a few hundreds, thus increasing enormously the total number of halo white dwarf candidates, which actually is of the order 10, in the best of the cases. However, the completitude of the sample even for 100 pc will be small — only of ~ 0.5. Most importantly, GAIA will only be able to observe the bright portion of the halo white dwarf luminosity function. Hence, a direct determination of the age of the halo using the cut-off of the halo white dwarf luminosity function will not be possible. Nevertheless, the bright portion of the halo white dwarf luminosity function is sensitive to the adopted IMF. Thus, GAIA will possibly constrain the precise shape of the IMF of the Galactic halo, if different from that of the disk. Despite all this, the accuracy of the measurements of those halo white dwarfs detected by GAIA will be impressive since we will have extremely precise parallaxes for a good fraction of halo white dwarfs with distances of up to 400 pc.

The natural question which arises now is how to distinguish halo white dwarfs from disk white dwarfs. Obviously, due to gravitational settling the metallicity cannot be used. Although GAIA will be able to obtain radial velocities using the SPECTRO instrument it is unlikely that GAIA could determine the full three-dimensional velocities of white dwarfs, since the SPECTRO instrument will be optimized for main sequence stars. The reduced proper motion diagram combined with advanced classification methods (Torres et al. 1998) can be of great help in distinguishing disk and halo members. An example of the expected reduced proper motion diagram that GAIA will obtain is shown in Figure 4. The reduced proper motion $H = M_{\rm V} - 5\log\pi + 5\log\mu$ is a good indicator of the membership to a given population. As it can be seen in Figure 4 halo white dwarfs occupy a clear locus in this diagram. However, for $V - I < 0.5$ the identification becomes less clear.

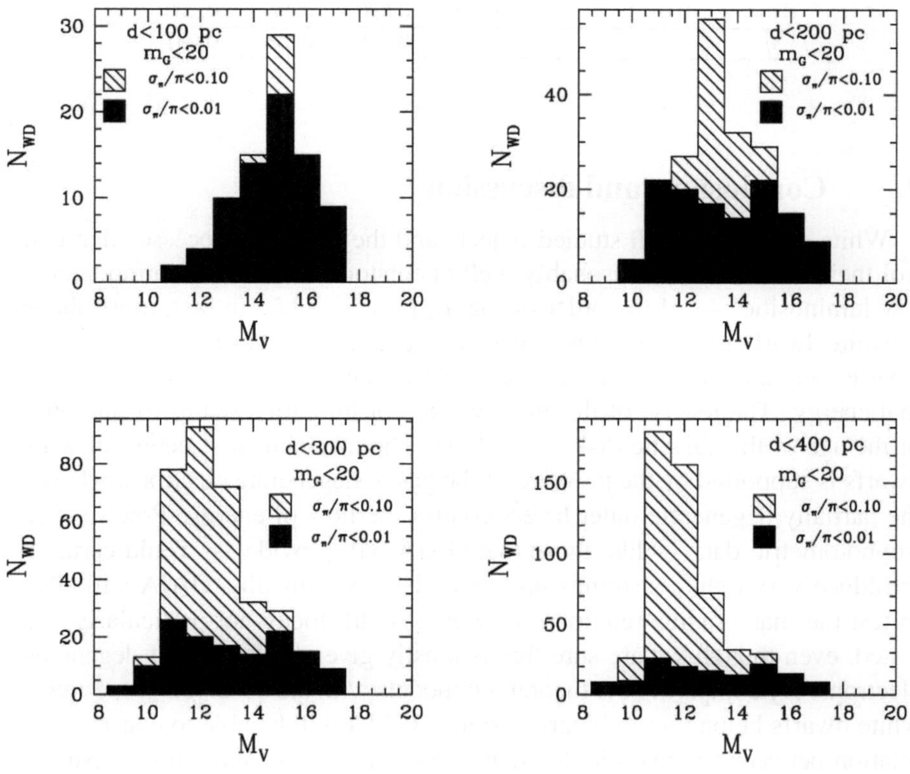

Figure 3. Same as Figure 2 for the halo white dwarf population.

Table 3. Same as table 2 for the halo white dwarf population.

	100 pc	200 pc	300 pc	400 pc
$N_{\rm WD}(9.0 < G < 29.0)$	359	2737	9174	21505
$N_{\rm WD}(G < 21)$	192	434	726	1099
$N_{\rm WD}(\mu > \mu_{\rm cut})$	192	434	726	1099
$\sigma_\mu/\mu < 0.10$	1.000	1.000	1.000	1.000
$\sigma_\mu/\mu < 0.01$	1.000	1.000	0.998	0.995
$\sigma_\pi/\pi < 0.10$	1.000	1.000	1.000	1.000
$\sigma_\pi/\pi < 0.01$	0.905	0.636	0.395	0.260
η	0.535	0.158	0.079	0.051

4. Conclusions and discussion

White dwarfs are well studied objects and the physical processes that control their evolution are reasonably well understood, at least up to moderately low luminosities — of the order of $\log(L/L_\odot) = -3.5$. In fact, most phases of white dwarf evolution can be succesfully characterized as a cooling process. That is, white dwarfs slowly radiate at the expense of the residual gravothermal energy. The release of this energy lasts for long time scales (of the order of the age of the galactic disk $\sim 10^{10}$ yr). The mechanical structure of white dwarfs is supported by the pressure of the gas of degenerate electrons, whereas the partially degenerate outer layers control the flow of energy. Precise spectrophotometric data — like those that GAIA will provide — would certainly introduce very tight constraints on the models. Specifically, GAIA will allow to test the mass–radius relationship, which is still today not particularly well tested, even though we are sure that is mostly given by the gas of degenerate electrons. By comparing the theoretical models with the observed properties of white dwarfs belonging to binary systems, GAIA will be able to constrain the relation between the mass in the main sequence and the mass of the resulting white dwarf. Moreover, GAIA will also provide very precise information on the physical mechanisms (crystallization, phase separation, ...) operating during the cooling process by comparing the theoretical luminosity functions of disk white dwarfs with the observations. Given their long cooling timescales, white dwarfs have been used as a tool to extract information about the past history of our Galaxy. The large number of white dwarfs that GAIA will observe will allow us to determine with unprecedented accuracy the age of the local neighborhood and the star formation history of the Galaxy. Furthermore GAIA will able to distinguish among the thin and the thick disk white dwarf populations, and, in this way, it will be able to provide a deep insight into the history of the galactic disk. It will also probe the structure and dynamics of

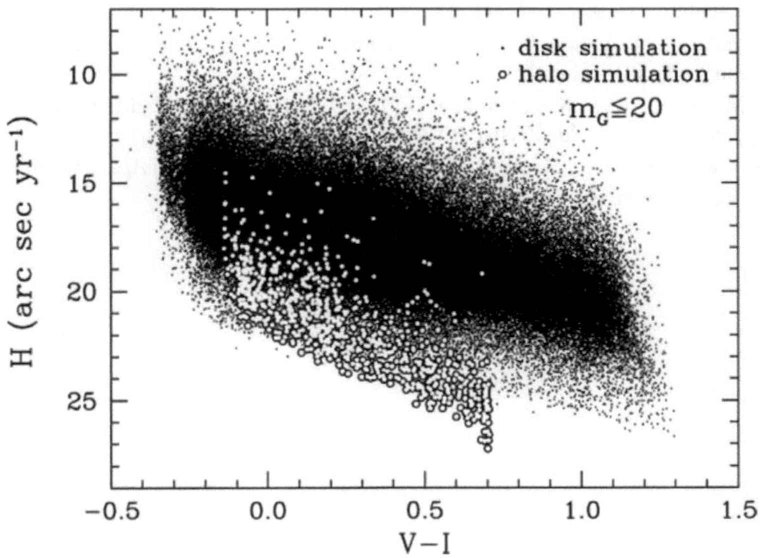

Figure 4. Reduced proper motion diagram for the disk — small solid dots — and halo — large open circles — simulations.

the Galaxy as a whole and it will provide new clues about the halo white dwarf population and its contribution to the mass budget of our Galaxy.

In summary, in this work we have shown how an astrometric mission like GAIA could dramatically increase the number of white dwarfs accessible to good quality observations. The increase of the observational database will undoubtely have a large impact in our current understanding of the history and structure of the Galaxy as well as on the theoretical models of white dwarf cooling, which, in turn, will for sure influence our knowledge of the physics of dense plasmas. Nevertheless, follow-up ground-based observations, theoretical improvements and accurate classification methods will be needed in order to analyze the disk and halo populations.

Acknowledgments

Part of this work was supported by the MCYT grants AYA04094–C03-01 and 02, and by the CIRIT.

References

García–Berro, E., Torres, S., Isern, J., & Burkert, A. 1999, *MNRAS*, **302**, 173.
García–Berro, E., Torres, S., Isern, J., & Burkert, A. 2004, *A&A*, **418**, 53.
Hakkila, J., Myers, J.M., Stidham, B.J., Hartmann, D.H. 1997, *AJ*, **114**, 2043.

Iben, I. Jr., Laughlin, G. 1989, *ApJ*, **341**, 312.

Perryman, M.A.C. 2002, in EAS Pub. Ser., Vol. 2, Proc. of *"GAIA: A European Space Project"*, EDP Sciences: Les Ulis (France).

Perryman, M.A.C., de Boer, K.S., Gilmore, G., Hoeg, E., Lattanzi, M.G., Lindegren, L., Luri, X., Mignard, F., Pace, O., de Zeeuw, P.T. 2001, *A&A*, **369**, 339.

Salaris, M., García–Berro, E., Hernanz, M., Isern, J. & Saumon, D. 2000, *ApJ*, **544**, 1036.

Torres, S., García–Berro, E., & Isern, J. 1998, *ApJ*, **508**, L71.

INFLUENCE OF METALLICITY IN THE DETERMINATION OF THE AGE OF HALO WHITE DWARFS

J. Isern[1,2], E. Garcia-Berro[1,3], M. Salaris[4], I. Dominguez[5]

[1]*Institute for Space Studies of Catalonia, Edifici Nexus, c/Gran Capità 2, 08034 Barcelona, Spain*

[2]*Institut de Ciències de l'Espai (CSIC)*

[3]*Departament de Física Aplicada, Universitat Politècnica de Catalunya*

[4]*Astrophysics Research Institute, Liverpool John Moores University, Twelve Quays-House, Egerton Wharf, Birkenhead CH41 1LD, UK*

[5]*Departmento de Física Teórica y del Cosmos, Universidad de Granada, Campus de Fuentenueva, 18071 Granada, Spain*

Abstract The authors study the effect of metallicity on post-AGB evolution and on the properties of remnant cores, and they examine the implications for distinct white dwarf populations in the Galactic disk and in the halo.

Keywords: Galactic disk, halo, stellar evolution

1. Introduction

One of the most interesting problems posed in modern astrophysics is the question of the formation of the galactic halo. Did it formed as a consequence of the rapid monolithic collapse of the protogalactic gas (Eggen et al. 1962) or did it formed through the accretion of fragments of tidally disrupted satellite galaxies (Searle & Zinn 1978)? Probably the answer is between these two extreme scenarios but still remains to be tested.

Since white dwarfs are the remnants of low–mass and intermediate–mass stars with cooling times of the order of the age of the Galaxy, it is natural to wonder if they can provide more insight to the problem. In principle, halo white dwarfs can be divided into three categories: i) those that formed during the monolithic collapse, ii) those that were accreted during the merging of satellite galaxies and iii) those that were expelled from the disc by the disruption of binaries or some other mechanism. These differences in the origin of

E.M. Sion, S. Vennes and H.L. Shipman (eds.), White Dwarfs: Cosmological and Galactic Probes, 25-30.
© 2005 *Springer*.

Table 1. Mass of the white dwarf that it is left by a one solar mass star and lifetime of the progenitor.

Z	$M_{\rm WD}^{H}$	$t_{\rm P}^{H}$(Gyr)	$M_{\rm WD}^{S}$	$t_{\rm P}^{S}$(Gyr)
0.0001	0.606	6.304	0.55	6.840
0.001	0.578	6.731	0.56	7.440
0.01	0.536	10.302	0.55	11.052
0.02	0.520	12.463	0.54	12.590

white dwarfs poses the problem of distinguishing among them. Since white dwarfs have lost all the information about their original metallicity, the only way to classify them is through their kinematics and age.

It is thus clear that the determination of the age of individual white dwarfs is critical and this depends not only on the cooling time but also on the lifetime in the main sequence. It is well known that low–Z stars have shorter lifetimes and produce larger degenerate cores than the corresponding high–Z stars of the same mass. Furthermore, metal deficient stars produce degenerate cores with less oxygen in the central regions which, because of the larger specific heat, reduces the cooling rate. Since we do not know the metallicity of the parent star, it is natural to ask ourselves about the influence of this parameter in the calculation of the age of a given individual white dwarf.

2. Stellar ages and mass of degenerate cores as a function of the metallicity

Recent evolutionary calculations have confirmed that, for the same initial mass, the lifetime of a star decreases with metallicity while the mass of the white dwarf that eventually is left shows the opposite behavior. This behavior can be easily understood taking into account that low values of Z induce a reduction of the opacity without a reduction of the radius. Therefore, in order to maintain the structure, the temperature and the luminosity have to increase with the subsequent reduction in the lifetime of the star. At the same time, the helium core that is left is larger and the white dwarf that finally results is more massive than those obtained from progenitors with the same mass but larger metallicities (Schwarzschild, 1958). The relationship between the mass of a white dwarf and that of its main-sequence progenitor is more controversial because of its complicated dependence on the way as different algorithms for handling convection, instabilities like breathing pulses and mass losses are used.

The aforementioned trends clearly appear in Table 1 where the time needed for a 1 $M_{\odot}$ to form a white dwarf is shown for different metallicities as computed by Hurley et al. (2000) and Salaris et al. (1997), t_P^H and t_P^S respectively.

Table 2. Temperatures, mass and mass uncertainty of suspectecd halo white dwarfs

Star	$\log(T_{\rm eff})$	$M_{\rm WD}$	$\Delta M_{\rm WD}$
WD 0145-174	3.8893	0.6	0.1
WD 0252-350	4.2319	0.37	0.02
WD 1022+009	3.7284	0.9	0.08
WD 1042+593	3.9212	0.6	0.1
WD 1448+077	4.1601	0.45	0.01
WD 1524-749	4.3695	0.46	0.01
WD 1756+827	3.8615	0.58	0.04
WD 2316-064	3.6767	0.68	0.11
WD 2351-368	4.1634	0.51	0.01

Table 1 also displays the corresponding mass of the white dwarfs that are left, $M_{\rm WD}^{H}$ and $M_{\rm WD}^{S}$. As it can be seen, the discrepancies, specially those referring to the mass of the white dwarf, are large.

3. The age of the white dwarf

Since stars form from a gas that has been progressively enriched with the metals ejected by previous generations, and they contribute themselves to the galactic metal content, it is natural to expect that young stars have a higher metal content than the older ones. The search for a definite relationship between the age and the metallicity has been the object of important efforts during the last twenty years. Although this a hot discussion topic, Twarog (1980), Carlberg et al. (1985), Meusinger et al (1991) and Rocha–Pinto et al. (2000) have provided good arguments in favor of a good correlation between age and metallicity. A simple approach that is enough for our purposes can be obtained from Figure 1 of Bravo et al. (1993). The conversion from metal abundance, [Me/H], to metallicity, Z, can be done as (Pols et al 1998):

$$Z = \frac{0.76}{3.0 + 37.425 \times 10^{[\rm Me/H]}}$$

where [Me/H]= log(Me/H)−log(Me/H)$_\odot$.

Table 2 shows the effective temperatures, masses and mass uncertainties for some suspected halo white dwarfs. Data were obtained from Bergeron (2003), Fontaine et al (2001), Leggett, Ruiz, & Bergeron (1998) and Pauli et al. (2003). It is interesting to note the existence in this table of four hot low–mass white dwarfs, which is not strange if it is accepted that the IMF is a function that has not changed much with either time or location. That is the same to say that the Galactic halo is still producing white dwarfs.

Figure 1 displays the isochrones obtained with the evolutionary sequences of Hurley et al. (2000) and Salaris et al. (1997). As it can be seen, the esti-

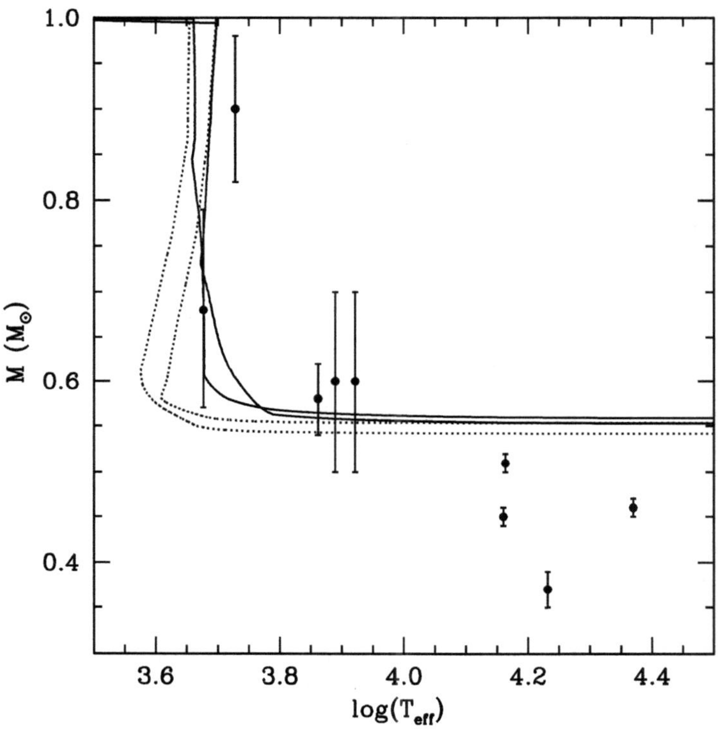

Figure 1. Theoretical isochrones corresponding to 13 and 11 Gyr obtained with the evolutionary tracks of Hurley et al. (2000) and Salaris et al. (1997) and the age metallicity relationship of Bravo et al. (1993) — solid and dotted lines respectively. See text for additional details.

mates of the ages of the coolest white dwarfs strongly depend on the adopted evolutionary tracks. For instance, in the case in which the evolutionary sequences of Salaris et al. (1997) are adopted, the coolest stars are too young to be members of the primordial halo — assuming that its formation started 13 Gyr ago and lasted less than 2 Gyr. On the contrary, in the case in which the theoretical isochrones of Hurley et al. (2000) are adopted, only WD 1022+009 is too young to be a member of the halo. At this point it is worthwhile to make two remarks:

- The cooling sequences were computed assuming that the mass of the hydrogen envelope is $M_{\rm H} = 10^{-4}\, M_{\odot}$ and we know from the distribution DA/non–DA ratio with the effective temperature that this figure can be different for individuals. Thus, a slight change in this figure can noticeably change the cooling rate of a white dwarf.
- In the case of the hottest white dwarfs, their mass is too small to be fitted with the models used here. This opens three possibilities: i) there is a systematic error in the determination of their mass, ii) models systematically overestimate the mass of white dwarfs produced by low–mass stars, and iii) all these stars are helium white dwarfs that are or were members of a close binary system.

4. Conclusions

We have presented a preliminary analysis of the influence of the metallicity in the determination of the age of white dwarfs. We have found that:

1 Metallicity plays a crucial role at the moment of assigning an age to an individual white dwarf (age of the progenitor plus cooling time).

2 The present uncertainties in the initial–final mass relationship do not allow to rule out any suspected halo white dwarf on the basis of the age alone.

3 If the IMF is universal, the halo should be still producing bright–low mass white dwarfs at present. However, the observed mass is much more smaller than the the computed mass of the putative members.

Acknowledgments

This work has been partially supported by the MCyT grants AYA2002-04094-C03-01/02 and 03, by the EU through FEDER funds, and by the CIRIT.

References

Bergeron, P. 2003, *ApJ*, **586**, 201.

Bravo, E., Isern, J., & Canal, R. 1993, *A&A*, **270**, 288.

Carlberg, R.G., Dawson, P.C., Hsu, T., & VandenBerg, D.A. 1985, *ApJ*, **294**, 674.

Eggen, D.I., Lynden–Bell, D., & Sandage, A.R. 1982, *ApJ*, **136**, 748.

Fontaine, G., Brassard, P., & Bergeron, P. 2001, *PASP*, **113**, 409.

Hurley, J.R., Pols, R.O., & Tout, C.A. 2000, *MNRAS*, **315**, 543.

Leggett, S.K., Ruiz, M.T., & Bergeron, P. 1998, *ApJ*, **497**, 294.

Meusinger, H., Reimann, H.–G., & Stecklum, A.A. 1991, *A&A*, **245**, 57.

Pauli, E.–M., Napiwotzki, R., Altmann, M., Heber, U., Odenkirchen, M., & Kerber, F. 2003, *A&A*, **400**, 877.

Rocha–Pinto, H.J., Maciel, W.J., Scalo, J., & Flynn, C. 2000, *A&A*, **358**, 850.

Salaris, M., Domínguez, I., Garcí–Berro, E. Hernanz, M., Isern, J., & Mochkovitch, R. 1997, *ApJ*, **486**, 413.

Schwarzschild, M. 1958, *Structure and Evolution of the Stars*, Dover, p.140.

Searle, L., & Zinn, R. 1978, *ApJ*, **225**, 35.

Twarog, B.A. 1980, *ApJ*, **242**, 242.

COOL HALO WHITE DWARFS FROM GSCII

D. Carollo, A. Spagna, M.G. Lattanzi, R.L. Smart
INAF-Osservatorio Astronomico di Torino, I-10025 Pino Torinese
carollo@to.astro.it

B. McLean
Space Telescope Science Institute, 3700 San Martin Drive, Baltimore, MD 21218, USA
mclean@stsci.edu

S.T. Hodgkin
Cambridge Astronomical Survey Unit, Institute of Astronomy, Madingley Road, Cambridge, CB3 0HA, UK
sth@ast.cam.ac.uk

L. Terranegra
INAF-Osservatorio Astronomico di Capodimonte, Via Moiariello 16, I-80131 Napoli, Italy
terraneg@na.astro.it

Abstract Microlensing experiments have suggested that a significant part of the dark halo of the Milky Way could be composed of matter in form of massive compact halo objects (MACHOs). Cool ancient white dwarfs (WDs) are the natural candidates. Here we present a new survey of halo WDs and evaluate the local space density using a new method for the membership selection based on the kinematic properties of the stars. A comparison to a revaluation of the Oppenheimer et al. (2001) result is also provided. The local space density estimated from the two independent samples is about $\sim 10^{-5} M_{\odot} pc^{-3}$ and is consistent with the canonical local mass density of the halo WDs.

Keywords: Galactic halo, white dwarfs

1. Introduction

Understanding the nature of the Baryonic Dark Matter in the galactic halo is a very tricky subject of astrophysics. The most favored candidates are brown dwarfs, planets, ancient cool white dwarfs (WDs), neutron stars and primordial

E.M. Sion, S. Vennes and H.L. Shipman (eds.), White Dwarfs: Cosmological and Galactic Probes, 31-40.
© 2005 *Springer.*

black holes. All these objects are known as MACHOs (Massive Compact Halo Objects) and, in principle, could be detectable by their gravitational microlensing effect on field stars (Pacynski 1986). The two major experiments for the detection of microlensing events, MACHOs (Alcock et al. 2000) and EROS (Afonso et al. 2003), seem to indicate that $\sim 20\%$ of the dark halo is made of compact objects with mass of $0.5M_{\odot}$. Very old cool white dwarfs with a mean mass around $0.5M_{\odot}$ are the natural candidates to explain the MACHOs results, even thought this interpretation is still controversial because an excess of white dwarfs today yield to an overproduction of red dwarfs and Type II supernovae (Torres et al. 2001).

Recently, several surveys have been implemented in order to detect a significant number of Population II white dwarfs. The most extensive one is that of Oppenheimer et al. (2001) who claimed the discovery of 38 halo white dwarfs and found a space density of 2% of the local halo dark matter. Different authors challenged these results based on the evidence that the kinematics of that sample seem to be more consistent with the thick disk population (Reid, Sahu & Hawley 2001; Reylé et al. 2001, Torres et al. 2002, Flynn et al. 2003) and on the age estimated for these WDs (Hansen 2001, Bergeron 2003). The basic problem is to discriminate the halo from the thick disk white dwarfs on the basis of the kinematic properties of the sample. Here a new high proper motion survey is presented based on the material used for the construction of the GSCII (Guide Star Catalogue II) (McLean et al. 2000), along with a new methodology for the membership selection.

2. Search for nearby Halo White Dwarfs

The aim of this survey is to search for halo WDs using plate material from the GSCII in the northern hemisphere and improve the measurements of halo WD space density. Also, important information could be inferred on the age of the thick disk and the halo through the analysis of the luminosity function, and the WD models could be improved adding experimental points to the color magnitude diagram. The recent improvement in the theory of atmosphere of cool WDs with hydrogen atmosphere yields to a cooling sequence which shows a turn off toward blue. This effect is due to the collision-induced-absorption in the infrared (H_2 CIA) which produce a redistribution of the flux toward shorter wavelength, and is strongly present at temperature of $T_{\rm eff} < 4000K$ (see, e.g., Chabrier et al. 2000). Also, the cooling models predict an absolute magnitude of $M_V \sim 16$ and 17.3 for a 0.6 $M_{\odot}$ WD of 10 and 13 Gyr respectively. Objects with these magnitudes are observable only within a few tens of parsecs with photographic material like GSC2, where the rare Pop. II WDs are mixed with the plenty of disk WDs and (sub)dwarfs. Anyway, suitable selection criteria, based on magnitudes, colors and proper motions provide a targets sample with a very poor contamination.

3. Plate material, processing and selection criteria

Our survey covers an area of ~ 1100 square degrees, mostly located toward the North Galactic Pole (NGP). The material consists of Schmidt plates from the Northern photographic surveys (POSS-I, Quick V and POSS-II) carried out at the Palomar Observatory. The second epoch survey provides blue, red and near-infrared magnitude, while the V magnitudes come from the Quick V survey. The plates are selected at high galactic latitude ($b > 70°$) in order to avoid the crowding.

All plates were digitized at STScI utilizing modified PDS-type scanning machines (Laidler et al. 1996). The digital copies of the plates were initially analyzed by means of the pipeline used for the construction of the GSC2 (McLean et al. 2000). The software provides classification, position, and magnitude for each object by means of astrometric and photometric calibrations which utilized Tycho2 (Hog et al. 2000) and GSPC2 (Bucciarelli et al. 2001) as reference catalogs. Accuracies better than 0.1-0.2 arcsec in position and 0.15-0.2 mag in magnitude are generally attained. In order to detect high proper motion objects, we chose POSS-II plates (blue, red, infrared) with epoch difference of $\Delta t \sim$ 1-10 yr. Then, the object matching and proper motion evaluation was performed using the procedure described in Spagna et al. (1996).

White dwarfs candidates are selected by means of various criteria. The first step is to identify faint ($R_F > 16$) and fast moving stars ($0.3 < \mu < 2.5''$/yr), then the proper motion of each target is confirmed by visual inspection of POSS-I and POSS-II plates in order to reject false detections (e.g. mismatches and binaries). Finally, a cross-correlation with other catalogues (2MASS, LHS, NLTT) is performed, as well.

A very useful parameter for the selection of the targets is the reduced proper motion (RPM), $H = m + 5\log\mu - 5$. The RPM diagram, H_R vs. $(B_J - R_F)$, was adopted to identify faint objects with high proper motion and to separate disk and halo WDs from late type dwarfs and subdwarfs. The RPM diagram of our survey is shown in Figure 1. Here, the triangles represent the sample of GSCII white dwarfs; 80% of this has been already confirmed spectroscopically, while the remaining 20% awaits confirmation, expected at the end of the last spectroscopic run (January 2004).

Spectral analysis is required for a confirmation of the nature of the selected candidates. In particular, low resolution spectroscopy is suitable to recognize the spectral type and the main chemical composition of the stars. Most of the spectroscopic observations were carried out with the 3.5 m Telescopio Nazionale Galileo (TNG, La Palma) with a significant support of the 4.2 m

William Herschel Telescope (WHT, La Palma) and the 2.2 m Isaac Newton Telescope (INT, La Palma) (see Carollo et al. 2004 for details) .

4. Analysis of the sample

The spectroscopic follow-up provided spectra for 50 stars and the total number of confirmed white dwarfs is 35 in which 14 are without H_α line, including a few exotic objects like the coolest ever carbon rich WD (Carollo et al. 2002, Carollo et al. 2003). The remaining objects are M dwarfs and sub-dwarfs. Figure 2 shows some example of DA and DC white dwarfs (left) and a few peculiar objects discovered (right).

In order to derive the halo or thick disk membership of the confirmed WDs, we first evaluated the photometric parallaxes using the M_F vs. $B_J - R_F$ relation. The calibration was based on a sample of WDs with known trigonometric parallax (Bergeron et al. 2001) and adopting the color transformations BVR $\Rightarrow B_J R_F I_N$ used for the GSCII calibration. We computed distances with an uncertainty of 25-30 $\%$ including the cosmic dispersion ($\sim$ 0.46 mag) and photometric errors ($\sigma_J \approx \sigma_F \approx$0.15 mag) (see Carollo et al. 2004).

Membership: thick disk or halo population?

The kinematic analysis of the WD sample drawn from a proper motion survey, including the choice of an optimal criteria to reject the contaminant disk WDs and select the true halo WDs, is one of the most critical part of this kind of studies.

Here, we consider as *bona fide* halo WDs those objects whose kinematics is *not* consistent with the velocity distribution of the thick disk population with a 99$\%$ confidence level which allows a reliable identification of halo WDs, while limiting the contamination of thick disk objects.

We assumed that in the solar neighbourhood ($\sim$ 100 pc) the thick disk population can be approximated by an uniform local space density and by a 3D gaussian Schwarzschild distribution:

$$p(\bar{v}) = \frac{1}{(2\pi)^{3/2}\sigma_U\sigma_V\sigma_W} \exp\left[-\frac{U^2}{2\sigma_U^2} - \frac{(V-V_0)^2}{2\sigma_V^2} - \frac{W^2}{2\sigma_W^2}\right] \qquad (1)$$

where (U,V,W) are the galactic velocity components, V_0 indicates the rotation lag with respect to the LSR and (σ_U, σ_V and σ_W) the velocity dispersions. In our analysis the thick disk parameters evaluated by Soubiran et al. (2003) have been adopted.

Because the radial velocity is not known, the full 3D space velocity cannot be recovered, therefore it is convenient to project Eq. 1 onto the tangential

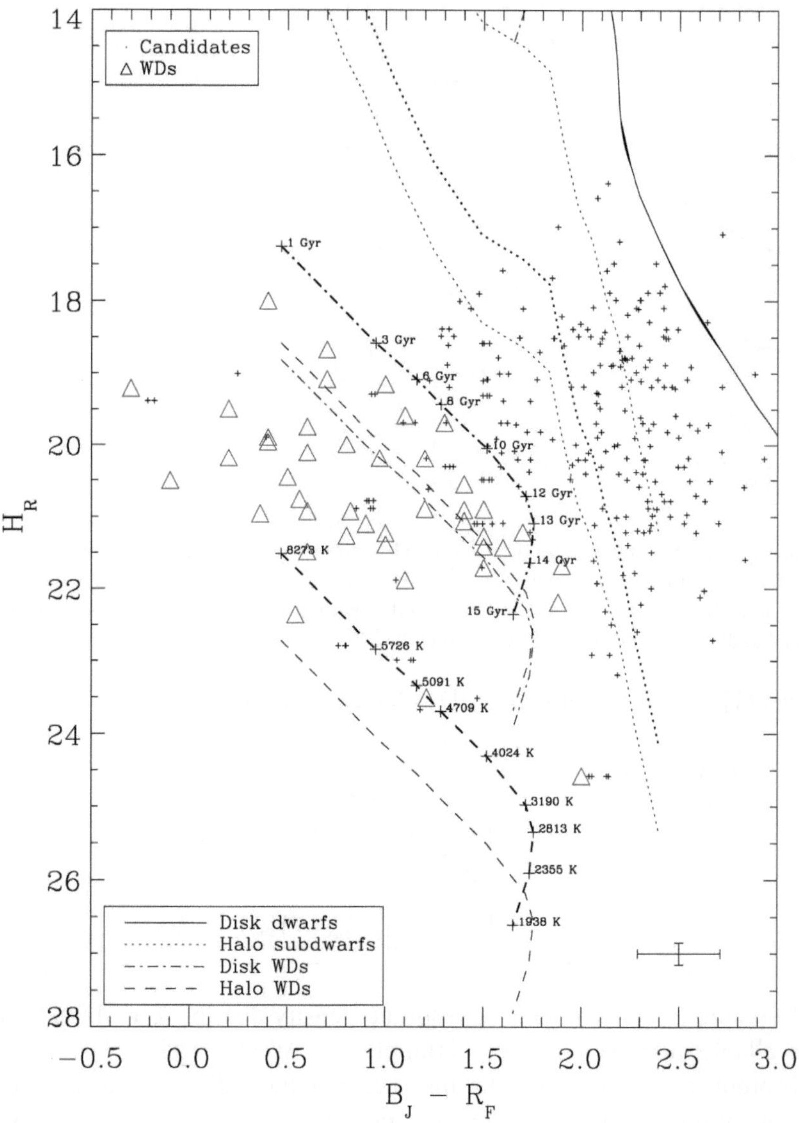

Figure 1. Reduce proper motion H_F versus $B_J - R_F$ of the candidates selected in an area of 1100 sq-deg. The thick solid and dotted lines show the locus of the disk dwarfs and the halo subdwarfs based on the 10 Gyr isochrones down to $0.08 M\odot$ from Baraffe et al. (1997, 1998) with [Fe/H]=0 and -1.5, respectively. The thick dashed and dot-dashed lines show the cooling tracks of $0.6 M\odot$ WDs with hydrogen atmosphere from Chabrier et al. (2000) of the halo and thin disc, respectively. We adopted mean tangential velocities (towards the NGP) of $V_T = 38$ km/s (disk) and 270 km/s (halo). Thin dashed and dot-dashed lines indicates the 2σ kinematics thresholds.

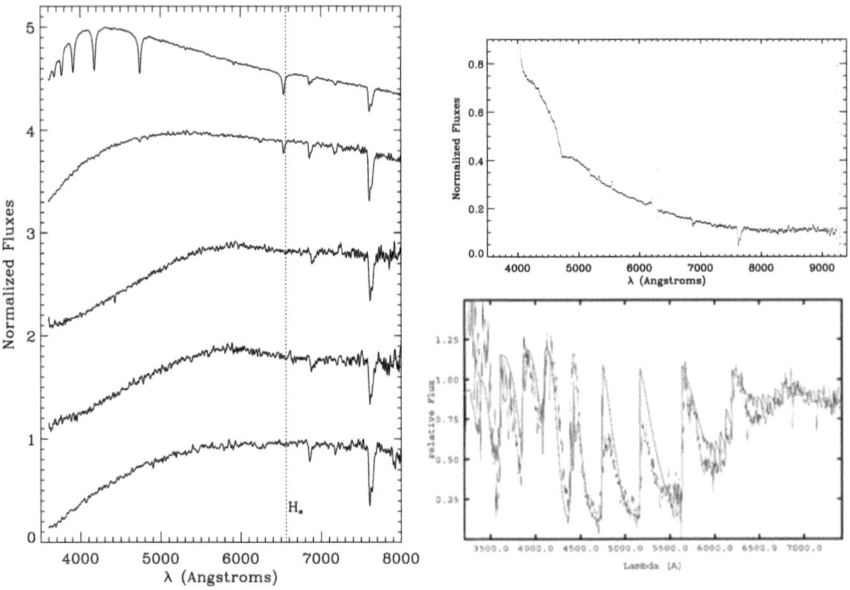

Figure 2. Left panel: spectra of new cool DA and DC WDs; the vertical dotted line indicates the H_α line. In the right panels two extreme peculiar objects: a very hot magnetic WD (top right) and the peculiar DQ WD (bottom right) with strong C_2 Deslandres-d' Anzabuja and Swan bands (Carollo et al. 2002; Carollo et al. 2003).

plane, (V_α, V_δ), and define the 2D velocity distribution:

$$\psi(V_\alpha, V_\delta) = \frac{1}{2\pi\sigma_\alpha\sigma_\delta\sqrt{1-\rho 2}} \exp\left[\frac{-1}{2(1-\rho^2)}\left(\frac{(V_\alpha - V_{\alpha 0})^2}{\sigma_{V\alpha}^2} - 2\rho\frac{(V_\alpha - V_{\alpha 0})}{\sigma_{V\alpha}}\frac{(V_\delta - V_{\delta 0})}{\sigma_{V\delta}} + \frac{(V_\delta - V_{\delta 0})^2}{\sigma_{V\delta}^2}\right)\right] \quad (2)$$

Equation 2 represents the appropriate density distribution which handle the unavailable radial velocity (cfr. Trumpler and Weaver 1953).

Moreover, we need to take into account the following observation constraints of the GSCII survey: (1) the apparent magnitude limit $R_F < R_{Flim} \simeq 19.5$ which implies a distance limit as a function of the absolute magnitude, M_F, of the WD ($r < r_{\max}(M) = 10^{[0.2(R_{Flim} - M_F)+1]}$), and (2) a proper motion limit $\mu > \mu_{lim} = 0.3"yr^{-1}$ which defines a tangential velocity threshold as a function of the distance ($V_{\tan} > V_{\min}(r) = 4.74\mu_{\lim}r$).

At this stage, as described in more details in Spagna et al. (2004), we can define the probability to find a star in the range (r,r+dr), $(V_\alpha, V_\alpha + dV_\alpha)$, $(V_\delta, V_\delta + dV_\delta)$ as $f(r, V_\alpha, V_\delta)drdV_\alpha dV_\delta$. Also, integrating over r the joint probability density $f(r, V_\alpha, V_\delta)$ we can evaluate the marginal density distribution, $h(V_\alpha, V_\delta)$, which quantifies the probability that an object with tangential velocities (V_α, V_δ) can be found somewhere *within* the whole volume $\frac{1}{3}\Omega\, r^3_{\rm max}$, where an object with absolute magnitude, M, could in principle be observed. At the same time, we can introduce the conditional probability, $t(V_\alpha, V_\delta|r)$, that an object with tangential velocities (V_α, V_δ) can be found *at* the measured distance.

Both the marginal distribution $h(V_\alpha, V_\delta)$ and the conditional probability, $t(V_\alpha, V_\delta|r)$, have been adopted to analyze our WD sample. Basically, the main effect of a proper motion limited survey is to undersample the low velocity objects and bias the complete distribution $\psi(V_\alpha, V_\delta)$, so that the probability density is redistributed from the low velocity regions towards the high velocity tails.

In Figure 3 (top left) the ellipses show the iso-probability contours of the tangential velocity distribution expected for thick disk stars, $\psi(V_\alpha, V_\delta)$, evaluated in the direction of one of the stars in our sample (GSC2_678_1), whose tangential velocity is marked with a filled circle. Here, the small dots represent a Montecarlo realization drawn from the proper motion limited distribution $h(V_\alpha, V_\delta)$ with a total of 2000 points. Note that GSC2_678_1 is located outside the 3σ contour so that, according to the complete distribution, it should be rejected as a thick disk stars with a confidence lever higher than 99%. On the other hand, if we consider GSC2_678_1 in a statistical sense, as a member of a kinematically selected sample, such conclusion may change as shown for instance in the top right of the same figure, which presents the marginal distribution $h(V_\alpha, V_\delta)$. In fact, in these case the star is located *within* the iso-probability contour delimiting the 99% confidence level so that it must be accepted as a thick disk star on the basis of the same criterium. Opposite is the case of WD GSC2_268_2 which falls outside the 99% iso-probability contour in both cases, and then it is confirmed as halo star.

5. Evaluation of the local space density

The kinematic analysis and the membership selection were applied to the GSCII sample and the local space density evaluated by the $1/V_{\rm max}$ method (Schmidt 1968). Also, this membership selection was used on the Oppenheimer et al. (2001) sample. We found 2 halo WDs for the GSC2 sample[1]. The resulting local space density evaluated with the Schmidt method for the case of the marginal distribution $h(V_\alpha, V_\delta)$ is $(0.55 \pm 0.47) \cdot 10^{-5} M_\odot pc^{-3}$ and the same value is obtained for the conditional distribution $t(V_\alpha, V_\delta|r)$. Using the same methodology on the Oppenheimer sample yields to a similar value of the order of $\sim 10^{-5}$ as described in Spagna et al. (2004). These values are consistent with the local mass density of halo stars estimated by Gould et

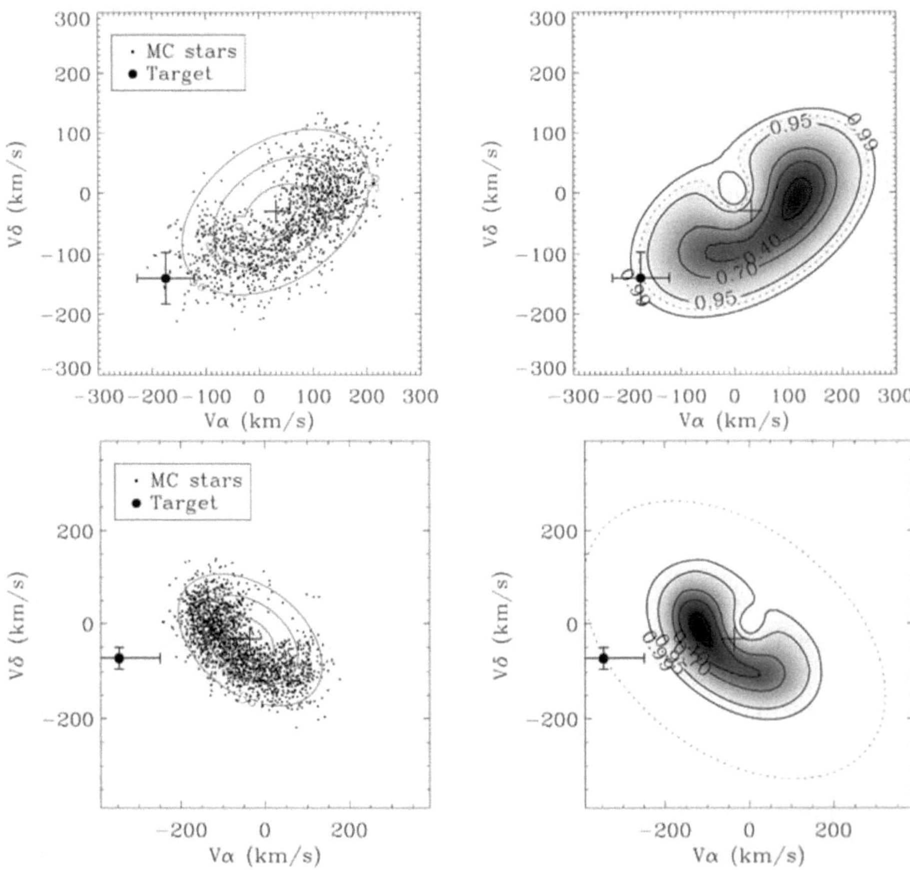

Figure 3. Examples of kinematic membership in the (V_α,V_δ) plane for two WDs (GSC2_678_1 and GSC2_268_2). In the left panels the solid lines represent the 1σ,2σ,3σ contours of the complete thick disk population $\psi(V_\alpha, V_\delta)$, while the dots show a Montecarlo simulations of a kinematically selected subsample ($\mu > 0.3''$ yr^{-1}, $r < r_{\rm max}$). In the right panels the solid lines show the iso-probability contours inferred from the marginal distribution $h(V_\alpha, V_\delta)$ with confidence levels up to 40% 70%, 95% and 99%, respectively. The WD position in the tangential velocity plane is represented by the filled circle. The dotted line shows the iso-probability contour corresponding to the WD velocity.

al. (1998), as well as by other authors who reanalyzed the Oppenheimer et al. sample (e.g. Reid et al. 2001, Reylé et al. 2001, Torres et al. 2002).

Acknowledgments

The GSC II is a joint project of the Space Telescope Science Institute and the Osservatorio Astronomico di Torino. Space Telescope Science Institute is operated by AURA for NASA under contract NAS5-26555. Partial financial support to this research comes from the Italian CNAA and the Italian Ministry of Research (MIUR) through the COFIN-2001 program.

This work is based on observations made with the Italian Telescopio Nazionale Galileo (TNG) operated on the island of La Palma by the Centro Galileo Galilei of the INAF (Istituto Nazionale di Astrofisica) at the Spanish Observatorio del Roque de los Muchachos of the Instituto de Astrofisica de Canarias, and also based on observations made with the William Herschel Telescope and the Isaac Newton Telescope operated on the island of La Palma by the Isaac Newton Group in the Spanish Observatorio del Roque de los Muchachos of the Instituto de Astrofisica de Canarias.

Notes

1. This is a preliminary result based on the subsample of WDs already confirmed which is the 80% of the whole sample. Spectroscopy of the remaining 20% will be available at the end of January 2004.

References

Afonso, C. et al. 2003, *A&A*, **404**, 145.
Alcock, C. et al. 2000, *ApJ*, **542**, 281.
Baraffe, I. et al. 1997, *A&A*, **327**, 1054.
Baraffe, I. et al. 1998, *A&A*, **337**, 403.
Bergeron, P. et al. 1997, *ApJ*, **108**, 339.
Bergeron, P. 2003, *ApJ*, **586**, 201.
Bucciarelli, B. et al. 2001, *A&A*, **368**, 335.
Carollo, D. et al. 2002, *A&A*, **393**, L45.
Carollo, D. et al. 2003, *A&A*, **400**, L13.
Carollo, D. et al. 2004, in preparation.
Chabrier, G. et al. 2000, *ApJ*, **543**, 216.
Flynn C. et al. 2003, *MNRAS*, **339**, 817.
Gould, A. et al. 1998, *ApJ*, **503**, 798.
Hansen, B. 2001, *ApJ*, **558**, 39.
Høg, E. et al. 2000, *A&A*, **355**, 27.
McLean, B. et al. 2000, *ASP Conf. Ser.*, **216**, 145.
Laidler, V.G. et al. 1996, *BAAS*, **188**, 54.21.
Luyten, W.J. 1979, NLTT, Minneapolis, Univ. of Minnesota.
Oppenheimer, B.R. et al. 2001, *Science*, **292**, 698.
Pacynski, B. 1986, *ApJ*, **304**, 1.
Reid, N.I., Sahu, K.C., & Hawley, S.L. 2001, *ApJ*, **559**, 942.

Reylé, C., Robin, A.C., & Créze, M. 2001, *A&A*, **378**, L53.

Schmidt, M. 1968, *ApJ*, **151**, 393.

Spagna, A. et al. 1996, *A&A*, **311**, 758.

Spagna, A. et al. 2004, in preparation.

Soubiran et al. 2003, *A&A*, **398**, 141.

Torres, S. et al. 2002, *MNRAS*, **336**, 971.

Trumpler, R. J., & Weaver, H. F. 1953, Dover Books on Astronomy and Space Topics, New York: Dover Publications.

DARK HALO BARYONS NOT IN ANCIENT WHITE DWARFS?

C. Reylé[1], M. Crézé[2], V. Mohan[3], A. C. Robin[1], J.-C. Cuillandre[4], O. Le Fèvre[5], H. J. McCracken[5,6], Y. Mellier[7,8]

[1] *Observatoire de Besançon, BP1615, F-25010 Besançon Cedex, France, email: celine@obs-besancon.fr*

[2] *Université de Bretagne Sud, BP 573, F-56017 Vannes Cedex, France*

[3] *U.P. State Observatory, Manora Peak, Nainital, 263129 India*

[4] *Canada-France-Hawaii Telescope, 65-1238 Mamalahoa Highway, Kamuela, HI 96743*

[5] *Laboratoire d'Astrophysique de Marseille, Traverse du Siphon, 13376 Marseille Cedex 12, France*

[6] *Osservatorio Astronomico di Bologna, via Ranzani 1, 40127 Bologna, Italy*

[7] *Institut d'Astrophysique de Paris, 98 bis, boulevard Arago, 75014 Paris, France*

[8] *LERMA, 61 avenue de l'Observatoire, 75014 Paris, France*

Abstract While there is little doubt about the existence of extended massive dark halos around galaxies, the possible contribution of baryons in the form of compact stellar objects in the dark mass is still poorly constrained. Microlensing experiments support the hypothesis that part of the Milky Way dark halo might be made of MACHOs with masses in the range 0.5-0.8 solar mass. Ancient white dwarfs are generally considered as the most plausible candidates for such MACHOs.

We report the results of a proper motion survey of a 0.16 deg^2 field at three epochs at high galactic latitude, and 0.94 deg^2 at two epochs at intermediate galactic latitude (VIRMOS survey), using the CFH telescope. Both surveys are complete to I = 24. The colour and proper motion data are suitable to separate unambiguously halo white dwarfs, which are identified by belonging to a non rotating system. No candidates were found within the magnitude-proper motion range, where such objects can be safely discriminated from any standard population.

A robust range of scenarii for a galaxy halo, made of only 3 to 5% in mass of ancient white dwarfs, predicts the detection of 1 to 2 such objects in our survey, if the halo is at least 14 gigayears old (more if younger). The recent limit of 13.5 gigayears set by the WMAP experiment on the age of first stars forces one to disregard schemes implying an older halo. We conclude that our observations

E.M. Sion, S. Vennes and H.L. Shipman (eds.), White Dwarfs: Cosmological and Galactic Probes, 41-48.
© 2005 *Springer*.

rule out at the 95% significance level a fraction of ancient white dwarfs in the dark halo greater than 15 to 25%.

Keywords: Dark matter, halo, white dwarf

1. Introduction

The rotation curves of galaxies show that they contain more mass than can be found in the form of luminous matter. This mass is not concentrated in the discs but rather in extended halos surrounding the luminous part of galaxies. In the Milky Way, the existence of an unseen disc has been ruled out from a detailed analysis of a sample of A and F stars observed by HIPPARCOS (Crézé et al. 1998). The local density of the dark matter halo amounts to 9.9×10^{-3} $M_\odot$ pc^{-3} whereas the stellar halo local density is found to be 10^{-5} $M_\odot$ pc^{-3} (Robin et al. 2003).

Among baryonic candidates, several objects have been proposed to form such massive halos while escaping direct detection, such as faint red dwarfs, Jupiter-like objects, brown dwarfs, or white dwarfs. From a search for faint red stars in the Hubble deep field, Flynn et al. (1996) showed that they cannot account significantly for the dark halo matter. Jupiter-like objects and brown dwarfs are eliminated by microlensing experiments EROS (Aubourg et al. 1993) and MACHO (Alcock et al. 1997). Combining results from both experiments, Lasserre (2000) estimated the dark halo content to be less than 20% of 0.5 to 0.8 $M_\odot$ massive compact objects. Although these results exclude fully baryonic halos, they still support that part of the dark halo could be made of faint old white dwarfs. Chabrier (1999) showed that ad hoc star formation scenarii are then necessary to form an adequate number of such objects together with producing a realistic heavy elements enrichment.

Surprisingly, white dwarfs which had long enough to cool cease to become fainter and redder, but remain at a constant luminosity while turning back towards the blue (Hansen 1998, Saumon & Jacobson 1999). Search for blue objects with high proper motion in Schmidt plates have shown the evidence of a substantial number of white dwarfs (Ibata et al. 2000, Oppenheimer et al. 2001). However, these surveys based on photographic plates are limited in depth and do not probe very far away from the Galactic plane. A large fraction of old disc and thick disc white dwarfs are expected to be found within this distance, which are hardly distinguishable from eventual dark halo white dwarfs. The debate is still on to understand whether these white dwarfs belong to the thick disc or halo population (Hansen 2001, Koopmans & Blandford, 2001, Reid et al. 2001, Reylé et al. 2001, Méndez 2002, Silvestri et al. 2002, Torres et al. 2002, Flynn et al. 2003).

Ibata et al. 1999 performed a second epoch observation in the Hubble Deep Field, allowing to probe a region beyond more than 1 kpc above the Galactic plane where disc and thick disc contributions are negligible. While they derived a few candidates from this study, a third epoch of observations did not confirm any of them (Richer 2002). From the EROS 2 proper motion survey, Goldman et al. 2002 found no halo white dwarfs candidates. Recently, Nelson et al. (2002) have found 5 high proper motion objects in a second-epoch Wide Field Planetary Camera 2 image of the Groth-Westphal strip. If white dwarfs, they would contribute from 1.5 to 7% of the dark halo mass.

In the present work, we attempt to detect halo white dwarfs in two deep CCD surveys obtained at the Canada-France-Hawaii telescope (CFHT). The Galactic volume probed lays outside the thick-disc dominated region. Surveys and data analysis are presented in Sect. 2. In Sect. 3, we use a Galactic model to isolate a region in the space of observable parameters where only halo white dwarfs can contribute and deduce a constraint on the white dwarfs halo mass fraction.

2. Surveys description and data analysis

Both have been completed at CFHT in Mauna Kea (Hawaii). The first one consists in three observing runs from 1996 to 2000 of 0.16 deg^2 in Selected Area SA57 near the North Galactic Pole. Observations were obtained in the V, R and I bands, with the UH8K camera during the first two runs, and the CFH12K camera for the last run. The second one is part of the VIRMOS-VLT Deep Survey (Le Fèvre et al. 2003), for which prior imaging observations have been obtained at CFHT. It covers an useful region of 0.94 deg^2 at intermediate Galactic latitude ($l = 171.7°, b = -58.2°$), observed by the CFH12K camera in the V and I bands at two epochs. The completness limit is I=24.

The source extraction was performed using the SExtractor package (Bertin & Arnouts 1996). This package also provides an object classification, the stellarity (CLS) ranging from 0 for non stellar objects to 1 for starlike objects. This parameter shows that stars and galaxies are well separated up to magnitude I=22, but for fainter objects the separation is more difficult. At I $>$ 22 the number of galaxies starts to be larger than the expected number of stars. Then a small proportion of misclassified galaxies implies a large contamination of the stars by these galaxies. Numerical experiments can help testing our star detection efficiency. 1500 fake stellar objects built on the observed PSF are added to the original frames. Then images are reprocessed and analysed through the complete treatment chain. Figure 1 gives the detection efficiency of the fake stars versus I magnitude. Pluses gives the number of detected stars over the number of created stars in 0.2 magnitude intervals. A polynomial fit within these points gives the efficiency curve. The upper curve is obtained considering all stars detected, the middle curve considering stars having CLS > 0.4 only, the lower curve considering stars having CLS > 0.8 only. The following thresholds are adopted: all objects with I < 21.5 and CLS > 0.8 are kept, as well as those with I > 21.5 and CLS > 0.4. With this selection, together with a large proper motion selection, most galaxies are withdrawn while no more than a few percent of stars are lost. Beyond I=23.5 the detection efficiency shrinks drastically while the CLS index fails definitely.

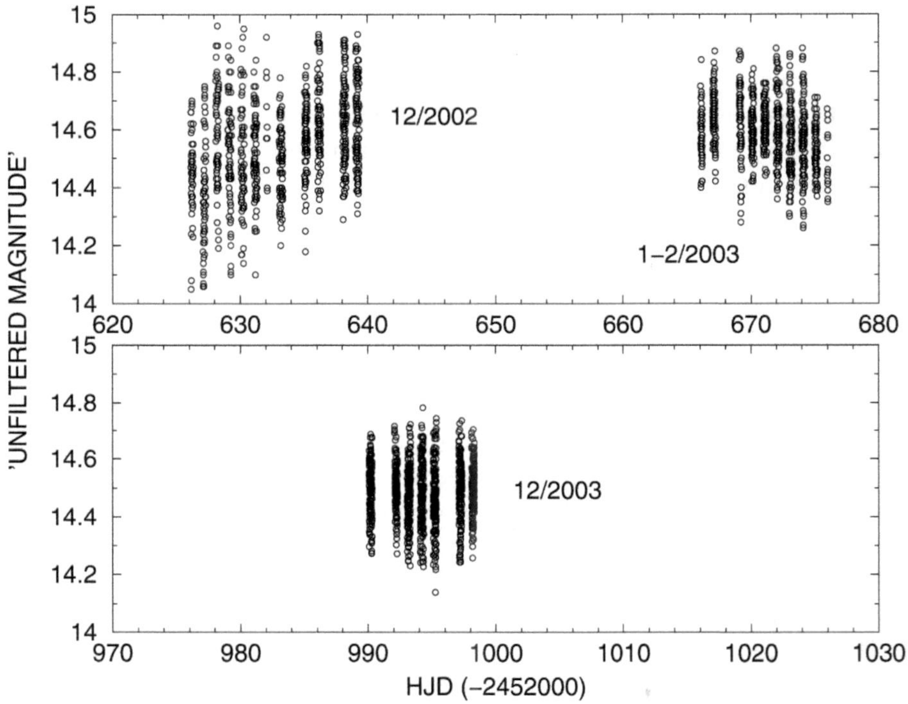

Figure 1. Detection efficiency curves versus I magnitude. Pluses gives the number of detected stars over the number of created stars in 0.2 magnitude intervals. A polynomial fit is superimposed. The upper curve is obtained with all detected stars, the middle curve with stars having CLS > 0.4, the lower curve with stars having CLS > 0.8.

Astrometry and photometry was performed using the DAOPHOT/ALLSTAR package (Stetson et al. 1998). The photometric accuracy for stellar objects is 0.02 for I < 22 and better than 0.1 for I = 24. Relative proper motions are obtained by cross-matching DAOPHOT positions separately within each CCD between two epochs and independently in the different bands. Standard proper motion errors for starlike objects is about $1''\text{cen}^{-1}$ at I = 21 and $7''\text{cen}^{-1}$ at I = 24 for the VIRMOS data. It is better than $1''\text{cen}^{-1}$ for the SA57 data.

3. Identification of halo white dwarfs

The upper panel in Figure 2 gives the proper motion μ_l of observed stars versus their I magnitude. The Besançon model of stellar population synthesis in the Milky Way is used to defined a region in the (μ_l,I) plane where halo white dwarfs are well separated from other populations.

The model features four populations (disc, thick disc, spheroid and bulge) each described by a star formation rate history, an initial mass function (IMF), an age or age-range, a set of evolutionary tracks, kinematics, metallicity characteristics, and includes a white dwarf population. The model is fully described in Robin et al. (2003) . In order to enhance the expected locus of halo white dwarfs in the (μ_l,I) diagram, we simulated the observed fields considering a white dwarf halo density of 9.9×10^{-2} $M_\odot$ pc^{-3}, that is 10 times higher than a dark halo fully made of white dwarfs. The kinematics of the dark halo is assumed identical with the one of the stellar halo: no rotation, ($\sigma_U/\sigma_V/\sigma_W$) = (131/106/85) km s^{-1} . The luminosity function has been adopted from Chabrier (1999) , who assumed two different IMFs. They are mainly constrained by the nucleosynthesis which places limits on the amount of high mass elements rejected in the interstellar medium at early ages. As there are not yet any direct observational constraints on the true IMF of these first stars, we have alternatively used both IMF to compute simulations. The IMF1 peaks at 3 $M_\odot$, and the mean white dwarf mass is 0.8 $M_\odot$. The IMF2 peaks at 2 $M_\odot$, and the white dwarfs are more luminous, with a mean mass of 0.7 $M_\odot$. The halo age is considered to be 14 Gyr.

The result of the simulation with IMF1 is shown on the lower panel of Figure 2. Halo white dwarfs appear with high proper motions. Using the simulated diagram, we defined a proper motion limit: it is 11 $''$cen^{-1} up to I = 21, and then increases due to the increase of proper motion errors with the magnitude. All objects beyond this limit are expected to be halo white dwarfs (except the one located at I = 19 and μ_l = 10 $''$cen^{-1} which is a dwarf of the old disc). On the observed diagram, objects are found to be above this limit. Their distribution in the (μ_l,I) plane is rather different from the one of simulated halo white dwarfs. All these objects were carefully examined and most of then where found to be artefacts, or extended objects. 3 objects, remaining possible candidates, have been observed with the VLT/FORS instrument, and have been found to be spurious moving objects.

Considering IMF1, the model predicted number of halo white dwarfs above the proper motion limit is 177. Given the detection efficiency, 126 objects are expected, that is 12.6 admitting a 100% white dwarf halo of 14 Gyr. Assuming a Poisson law, the zero detection is compatible with the expectation of 3 objects at the 95% confidence level. This sets an upper limit for the contribution of extreme white dwarfs in the mass of the dark halo of 24%. Considering IMF2,

275 simulated white dwarfs lie above the proper motion limit. 20.4 halo white dwarfs are expected in a 100% white dwarf halo of 14 Gyr. The upper limit to the halo mass fraction made of white dwarfs is then 15%.

4. Conclusion

This negative result combined with those of recent similar investigations, points at a rather low fraction of the massive halo in the form of stellar remnants, safely less than 25%. The recent limit of 13.5 gigayears set by the WMAP experiment on the age of first stars forces one to disregard schemes implying an older halo. On the contrary, to consider a more recent halo starburst would lower this fraction. Thus this result is in agreement with the limit for the MACHO density established by the EROS and MACHO Collaborations.

A more accurate estimation of the white dwarf halo mass fraction lies on deep surveys like the present one, depth where the thick disc ambiguity vanishes. Indeed, the combination of present data with data from HST (Nelson et al. 2002) and EROS2 (Goldman et al. 2002) drives the halo white dwarf mass fraction down to below 5% (Crézé et al., in preparation).

Furthermore, as part of the ongoing CFHT Legacy Survey, 1300 $\deg^{-2}$ of the sky will be observed twice over 6 years, allowing to retrieve proper motion objects. More than 300 white dwarfs with proper motion larger than 50 $''\mathrm{cen}^{-1}$, that is unambiguously belonging to the halo, are expected to be found in such survey with a halo of 14 Gyr if 10% made of white dwarfs.

Acknowledgments

Based on observations made at Canada France Hawaii Telescope (CFHT). This work was partially supported by the Indo-French Centre for the Promotion of Advanced Research (IFCPAR), New Delhi (India).

References

Alcock, C. et al. 1997, *ApJ*, **486**, 697.

Aubourg, E. et al. 1993, *Nature*, **365**, 623.

Bertin, E. & Arnouts, S. 1996, *A&A Suppl. Ser.*, **117**, 393.

Bienaymé, O., Robin, A. C., & Crézé, M. 1987, *A&A*, **180**, 94.

Chabrier, G. 1999, *ApJ*, **513**, L103.

Crézé, M., Chereul, E., Bienaymé, O., & Pichon, C. 1998, *A&A*, **329**, 920.

Flynn, C., Gould, A., & Bahcall, J. N. 1996, *ApJ*, **466**, L55.

Flynn, C., Holopainen, J., & Holmberg, J. 2003, *MNRAS*, **339**, 817.

Goldman, B. et al. 2002, *A&A*, **389**, L69.

Hansen, B. M. S. 1998, *Nature*, **394**, 860.

Hansen, B. M. S. 2001, *ApJ*, **558**, L39.

Ibata, R. A., Richer, H. B., Gilliland, R. L., & Scott, D. 1999, *ApJ*, **524**, L95.

Ibata, R., Irwin, M., Bienaymé, O., Scholz, R., & Guibert, J. 2000, *ApJ*, **532**, L41.

Koopmans, L. V. E., Blandford, R. D. 2001, *MNRAS*(astro-ph0107358).

Lasserre, T., 2000 , PhD Thesis.

Le Fèvre, O., Mellier, Y., McCracken, H.J., et al. 2004, *A&A*, **417**, 839.

Méndez, R. A. 2002, *A&A*, **395**, 779.

Nelson, C. A., Cook, K. H., Axelrod, T. S., Mould, J. R., & Alcock, C. 2002, *ApJ*, **573**, 644.

Oppenheimer, B. R., Hambly, N. C., Digby, A. P., Hodgkin, S. T., & Saumon, D. 2001, *Science*, **292**, 698.

Reid, I. N., Sahu, K. C., & Hawley, S. L. 2001, *ApJ*, **559**, 942.

Reylé, C., Robin, A. C., & Crézé, M. 2001, *A&A*, **378**, L53.

Richer, H. B. 2002, in "The Dark Universe: Matter, Energy and Gravity" (astro-ph0107079).

Robin, A. C., Reylé, C., Derrière, S., & Picaud, S. 2003, *A&A*, **409**, 523.

Saumon, D. & Jacobson, S. B. 1999, *ApJ*, **511**, L107.

Silvestri, N. M., Oswalt, T. D., & Hawley, S. L. 2002, *AJ*, **124**, 1118.

Stetson, P. B. et al. 1998, *ApJ*, **508**, 491.

Torres, S., García-Berro, E., Burkert, A., & Isern, J. 2002, *MNRAS*, **336**, 971.

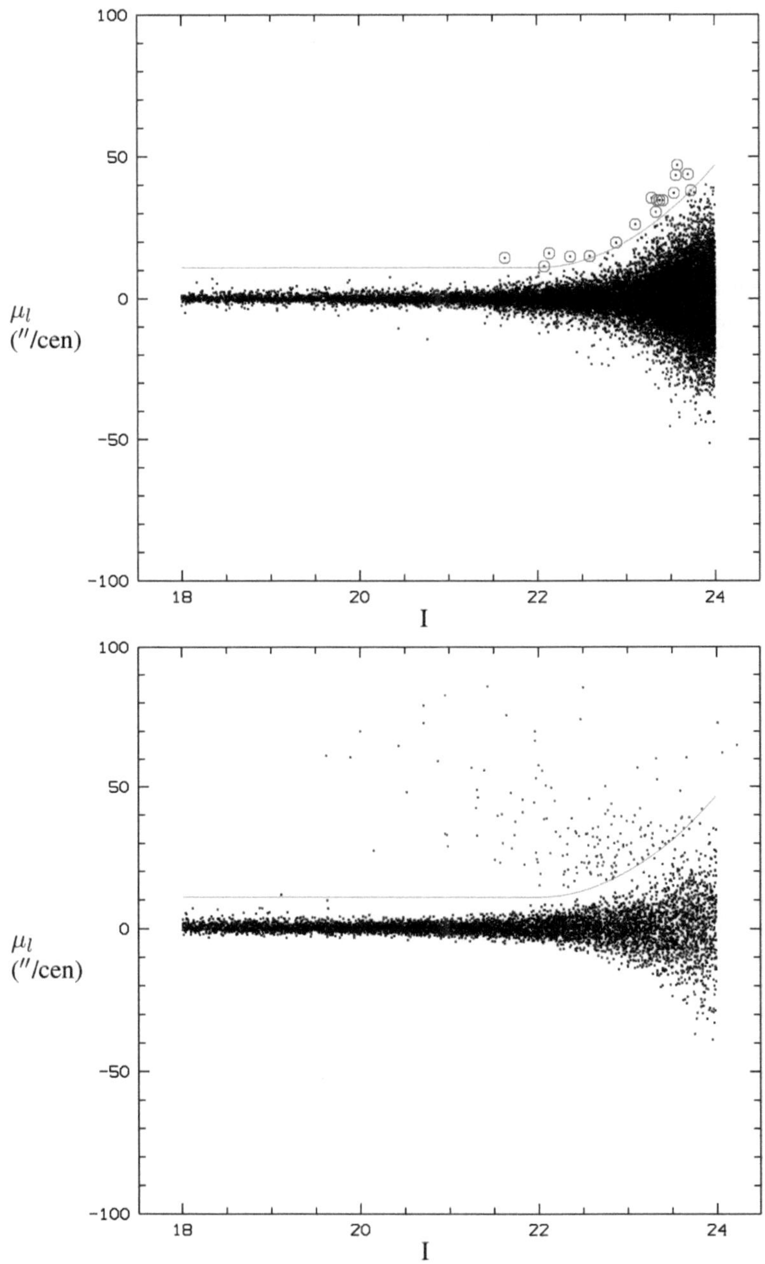

Figure 2. Magnitude/proper motion diagram for the VIRMOS data. Upper panel: observed stars. Lower panel: simulated stars considering a halo white dwarfs density 10 times greater than the one of a halo fully made of white dwarfs, IMF1 from Chabrier (1999), and a halo age of 14 Gyr. All simulated stars above the curve are halo white dwarfs. A carefull check of all observed objects above this limit (circled dots) shows they are non stellar.

THE WHITE DWARF POPULATION IN THE 2QZ AND SLOAN SURVEYS

Stéphane Vennes
Department of Physics and Astronomy, Johns Hopkins University
Baltimore, MD 21218-2686, USA; vennes@pha.jhu.edu

Adéla Kawka
Astronomický ústav AV ČR, Fričova 298, CZ-251 65, Ondřejov, Czech Republic

Scott M. Croom, Brian J. Boyle
Anglo-Australian Observatory, PO Box 296, Epping, NSW 1710, Australia

Robert J. Smith
Liverpool John Moores University
Twelve Quays House, Egerton Wharf, Nirkenhead CH41 1LD

Tom Shanks, Phil Outram
Department of Physics, University of Durham, South Road, Durham DH1 3LE

Lance Miller, Nicola Loaring
Department of Physics, Oxford University, 1 Keble Road, Oxford OX1 3RH

Abstract We present an analysis of the Anglo-Australian Telescope 2dF QSO Redshift Survey (2QZ) of high-Galactic latitude white dwarf stars. Approximately 2400 white dwarfs have been identified with distances above the Galactic plane ranging from 200 pc to 1 kpc. The same stellar population has been re-discovered in the Sloan Digital Sky Survey and we determine their temperature, mass, and absolute luminosity distributions with an analysis of the Sloan photometric and 2QZ spectroscopic data sets. Our main objectives are to measure the scale-height and luminosity function of the population, determine its formation history and examine evidence for spectral evolutionary effects.

Keywords: Milky Way, surveys, white dwarfs

E.M. Sion, S. Vennes and H.L. Shipman (eds.), White Dwarfs: Cosmological and Galactic Probes, 49-60.
© 2005 *Springer.*

1. Introduction

The white dwarf luminosity and mass functions describe properties of the population which constrain Galactic evolutionary models (e.g., Wood 1992). Recent surveys such as the 2dF QSO redshift survey (2QZ) and the Sloan Digital Sky Survey (SDSS) are assembling large homogeneous data sets which support a study of the Galactic population of white dwarf stars. Our main objectives are to map the distribution of white dwarfs in the Galaxy and measure its scale height, and, secondly to build luminosity functions for hydrogen-rich (DA) and helium-rich (non-DA) white dwarfs.

The 2QZ Survey is a deep spectroscopic survey of blue Galactic and extragalactic sources which was conducted at the Anglo-Australian Telescope in Australia. The first release of the 2QZ survey (Croom et al. 2001) resulted in the discovery of 942 new white dwarfs. Vennes et al. (2002) obtained atmospheric parameters for many of these objects, and, for the first time, observed directly the exponential fall-off of the population density as a function of height above the Galactic plane. The white dwarf population observed in the 2QZ survey belongs to the so-called thin-disk of the Galaxy and is characterized by a vertical scale-height of 300 pc compared to a scale-height of 1350 pc for the thick disk (Gilmore & Reid 1983; Chen et al. 2001). The final release of the 2QZ survey is now available and comprises $\sim$ 2400 white dwarfs (Croom et al. 2004).

The SDSS aims to cover a quarter of the sky with photometric colors of 200 million objects, both Galactic and extra-galactic, and calibrated spectroscopy for a substantial fraction of these objects. The data is available on the web through the *SkyServer Data Mining* interface (http://skyserver.pha.jhu.edu/), but a large spectroscopic catalog of 2551 new white dwarfs is already available (Kleinman et al. 2004).

We describe the selection process for the white dwarfs in the 2QZ and Sloan surveys, followed by a summary of the main results from the 2QZ survey— white dwarf scale-height and birthrate— and from the Sloan survey following the first release. We exploit the overlap area between 2QZ and Sloan surveys and extract Sloan photometry, and we derive a preliminary luminosity function for the joint 2QZ/Sloan population. Finally, we summarize our current efforts and propose future work.

2. The 2QZ survey: population density and scale height

The white dwarfs have been observed serendipitously as part of the Anglo-Australian spectroscopic survey using the 4-m telescope and the 2-degree Field multi-fiber spectrograph. The survey covers a nominal 722 deg^2, or an effective 673 deg^2 in the final release (Croom et al. 2004). The survey magnitude limits are $18.25 \leq b_J \leq 20.85$ (see Croom et al. 2001). The spectroscopic tar-

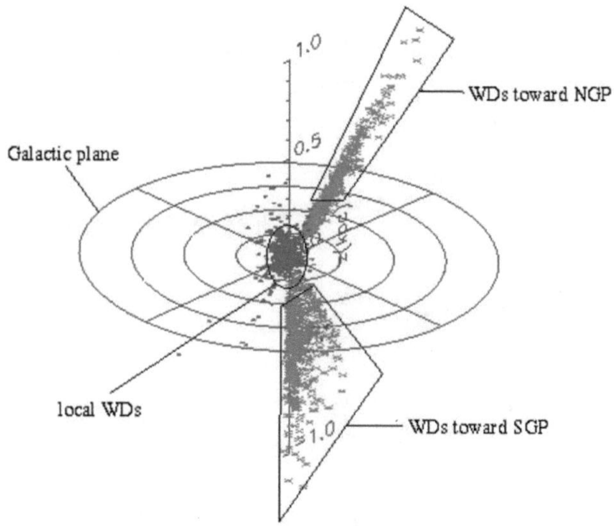

Figure 1. Spatial distribution of 2QZ selected white dwarfs toward the south and north Galactic poles (From Vennes 2003, 13th European Workshop on White Dwarfs). Local white dwarfs from the McCook & Sion (1999) catalog are included in the diagram.

gets are selected on 30 UK Schmidt Telescope plates searched for blue objects using the photographic indices $u - b_J$ and $b_J - r$. The survey selected 2074 DA white dwarfs (including 71 DA+dM) white dwarfs and 326 DB/DZ white dwarfs at high-Galactic latitudes toward the south Galactic pole (SGP) and the north Galactic pole (NGP). Figure 1 shows the spatial distribution of DA white dwarfs from the 2QZ 10k release (Croom et al. 2001)

Several spectra of white dwarfs, generally characterized by $T_{\rm eff} \geq 10,000$ K, have been obtained. Figure 2 shows a hot DA white dwarf with $T_{\rm eff} = 24,700$ K, and $\log g = 8.5$, a He-rich DZ white dwarf showing the CaII H&K doublet, a He-rich DQ white dwarf showing molecular carbon bands, and a DB white dwarf with $T_{\rm eff} \sim 16,000$ K.

Some basic results are obtained from a number count analysis of 1934 DA and 314 DB/DZ white dwarfs, pre-selected from the catalog in June 2003. Figure 3 shows the surface density per magnitude bins as a function of the photographic b_J magnitude for the two populations along with the Palomar-Green DA density (Fleming et al. 1986). The distributions are compared to an exponential distribution above the Galactic plane ($h = 275$ pc). The expected number densities are integrated assuming constant birthrates, $b_{DA} = 0.5 \times 10^{-12}$ and $b_{DB} = 0.7 \times 10^{-13}$ pc^{-3} yr^{-1}, close to the average birthrate over the lifetime of the disk, and assuming $M_V \leq 12.75$. Based on this rudimentary analysis, we may already conclude that the ratio of DA to non-DA is

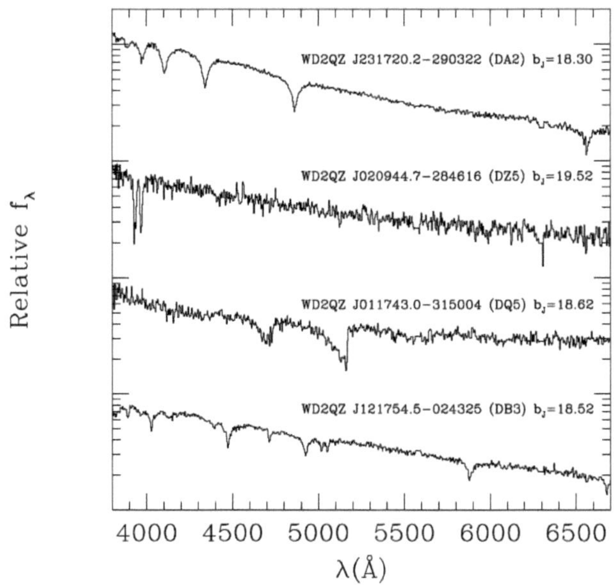

Figure 2. Example of spectra from four important classes of white dwarfs, the hydrogen-rich DA, the helium-rich DZ and DQ contaminated with calcium and carbon, respectively, and the hot helium-rich DB white dwarfs. Compare these spectra with the spectroscopic atlas of Wesemael et al. (1993).

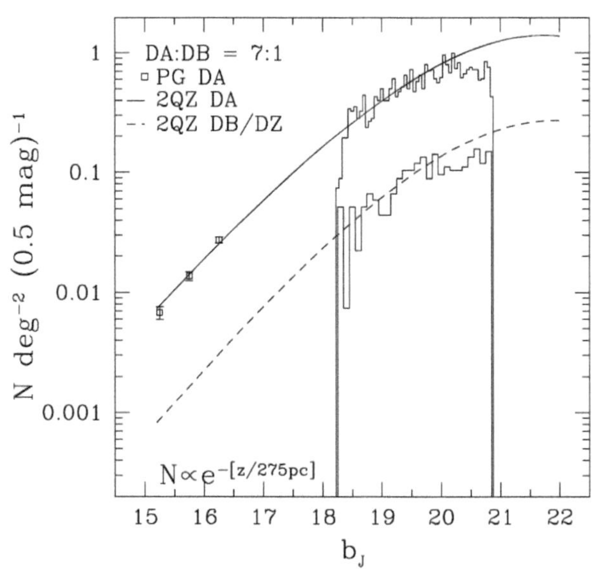

Figure 3. Number of white dwarfs per square degrees per half-magnitude bins divided in two broad spectral classes, the hydrogen-rich DA and the helium-rich DB/DZ white dwarfs, and compared to exponential models.

$n(\mathrm{DA})/n(\mathrm{non-DA}) \sim 7$, and that the total white dwarf birthrate (DA+DB) is close to 0.6×10^{-12} pc^{-3} yr^{-1}.

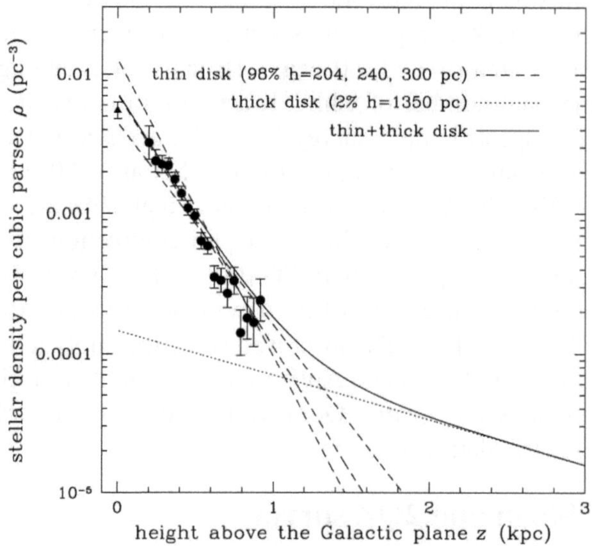

Figure 4. Number density of DA white dwarfs in the 2QZ 10k release as a function of height above the Galactic plane and normalized to the local space density (From Vennes 2003, 13th European Workshop on White Dwarfs).

Figure 4 compares the observed DA white dwarf spatial distribution to model distributions with scale-heights of 204, 240, and 306 pc. The measured stellar density above the Galactic plane has been determined using a V/V_{max} method. The observed distribution has been renormalized to the local space density (DA *plus* non-DA; Holberg et al. 2001). The models are fitted to the overall distribution, assuming that 98% of all white dwarfs observed locally belong to the thin disk and that only 2% belong to the thick disk. The exponential decline is observed to a height of 1 kpc, but the effect of the thick disk would become apparent only at a height of at least 1.5 kpc above the plane.

3. Sloan: a spectroscopic and photometric database

The Sloan survey will cover 10,000 deg^2 of the north Galactic cap, and will offer a very comprehensive picture of the Galactic white dwarf population, its kinematics as well as its luminosity and mass functions. The Sloan photometry covers the optical spectrum in five broad bands (u, g, r, i, z) from 3540 Å to 9049 Å, and the spectroscopy covers approximately the same range.

Following the early release of the Sloan Digital Sky Survey, Harris et al. (2003) estimates the density of white dwarfs in the photometric catalog at ~ 2.2 deg^{-2}. They also reported the spectroscopic identification of 269 white dwarfs over an area of 190 deg^2, where only 44 were previously known. Their original assessment was improved following the first release of the Sloan survey. Kleinman et al. (2004) report the spectroscopic identification of 2551 white dwarfs over an area of $\sim$ 1360 deg^2. Because of changing target selection criteria, the final number of white dwarf spectra is uncertain but may exceed 15,000. The spectroscopic survey helps characterize the white dwarf population in great details. For example, Raymond et al. (2003) reported the identification 109 WD+dM binaries; it is expected that 10% are members of the pre-cataclysmic variables (CV) family. Upon completion of the survey, some 700-1000 pairs are expected with 70-100 of them pre-CVs. Schmidt et al. (2003) reported the discovery of 53 new magnetic white dwarfs, doubling the number known before. The improved statistics imply a link between high-field white dwarfs and magnetic Ap/Bp stars, but another origin is sought for low field magnetic white dwarfs. Our own investigation, described below, exploits the photometric database.

4. A joint Sloan and 2QZ survey

The Sloan First Data Release and the 2QZ Final Release coverages coincide over an equatorial band (Fig. 5). Spectroscopic data from the 2QZ survey are available for close to one thousand white dwarfs in this area, and spectroscopic and photometric data from the Sloan survey data are available for a fraction of these objects. The overlapping survey area is close to 40% of the final release of the 2QZ survey, but the southern Galactic cap region covered by 2QZ is beyond the reach of Sloan.

The Sloan database was searched for white dwarf candidates using the 2QZ white dwarf identifications: a search radius of 5 arcseconds was adopted allowing proper-motions of up to 200 mas per year. Some 711 DA candidates and 78 non-DA candidates were identified in the Sloan photometric database.

Figure 6 shows the measured color indices $u-g$ versus $g-r$ for the sample of 711 joint 2QZ-Sloan DA white dwarfs compared to synthetic colors, and Figure 7 shows the measured and calculated indices $g-r$ versus $r-i$.

The DA synthetic colors have been computed using synthetic spectra based on a grid of pure-hydrogen, convective model atmospheres, and 'ugriz' bandpasses defined on the AB system. The grid shows from right-to-left predicted colors at T_{eff} = 7, 10, 15, 20, 30, 40, and 60×10^3 K, and surface gravities, from bottom-to-top, at log g = 7.0, 7.5, 8.0, 8.5, 9.0, and 9.5. The main-sequence is also shown.

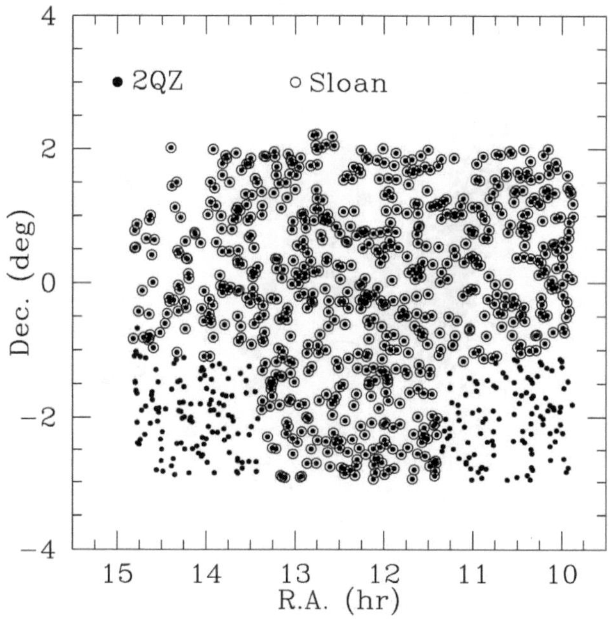

Figure 5. Overlap between 2QZ and Sloan surveys toward the north Galactic cap. With a few exceptions, all 2QZ sources have a counterpart in the Sloan photometric database.

Close to 85% of the joint 2QZ-Sloan white dwarfs have colors falling within the range of white dwarf stars (606 objects out of a selection of 711). A fraction of the remaining stars are suspected spectroscopic composites, but some clearly correspond to main-sequence colors, and are possibly field horizontal branch stars (halo). The DA white dwarf sequence appears extremely broad in both color-color diagrams, although a very narrow distribution is theoretically expected in the $g - r$ versus $r - i$ diagram. Therefore, this effect must be attributed to external causes, such as composite spectra in the red part of the spectrum, or variations in Galactic extinction in the blue. The effect cannot be attributed alone to intrinsic causes, although a broad white dwarf mass distribution may still dominate the spread in $u - g$ colors.

5. Mass and luminosity functions

Figure 8 presents the gravity-temperature diagram for the joint 2QZ-Sloan DA white dwarf sample, and the corresponding mass-age diagram. The temperatures and gravities were extracted from the $u - g$ and $g - i$ indices. Straddling the Balmer jump, the $u-g$ color index is most sensitive to surface gravity, while the extended $g - i$ index, combining the $g - r$ and $r - i$ indices, is most sensitive to effective temperature. Because of a considerable spread in col-

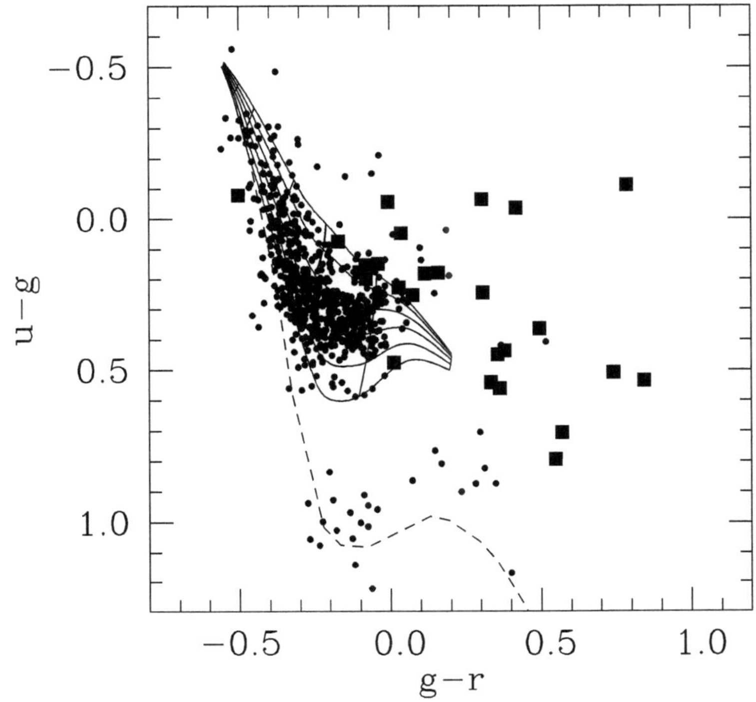

Figure 6. Color-color $u - g$ versus $g - r$ diagram showing the loci of hydrogen-rich DA white dwarfs (circles) and DA+dM pairs (squares) compared to predicted white dwarf colors (full lines) and the main-sequence (dashed line).

ors, the mass distribution appears very broad with a large number of low-mass ($M \leq 0.4$) and high-mass objects ($M \geq 0.8$).

Figure 9 compares the luminosity function of the joint Sloan-2QZ sample, the 2QZ luminosity function, the Palomar-Green luminosity function, computed using a $V/V_{\rm max}$ method applied to magnitude-limited samples, and theoretical expectations assuming a single mass (0.6) and a constant birthrate. The Palomar-Green and Sloan-2QZ surveys assume a scale-height of 250 pc, while Vennes et al. (2002) assumed a scale-height of 300 pc in the 2QZ survey alone.

Despite previously mentioned difficulties, the joint 2QZ-Sloan luminosity function appears complete from $M_V = 10$ to 12.5 owing to superior luminosity indicators for a much larger number of objects than available in the 2QZ survey alone. Fewer objects with $M_V \geq 12.5$ ($T \sim 9,000$ K) have been selected in the 2QZ survey due to the limiting colors at the cool end.

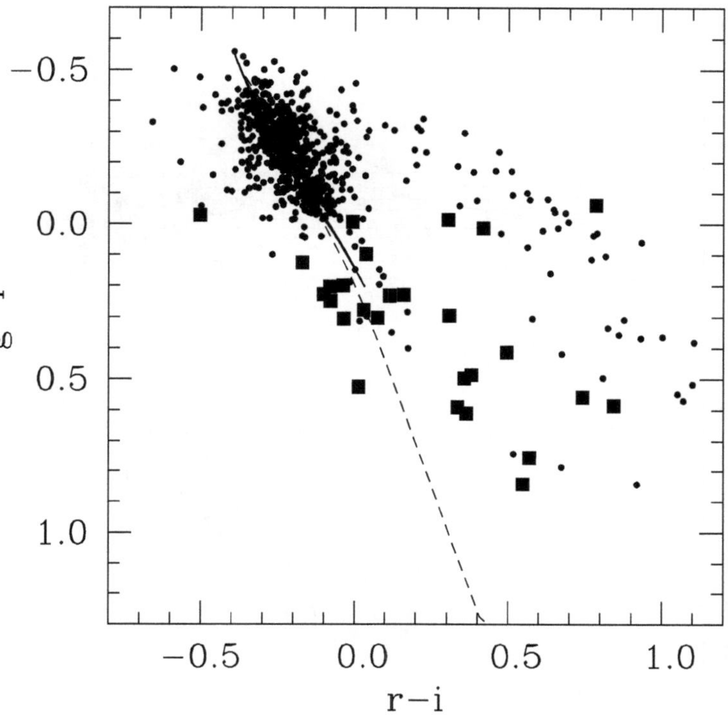

Figure 7. Same as Figure 6 but for the $g - r$ versus $r - i$ colors.

6. Conclusions

A study of a population of distant white dwarfs is under way using the final release of the 2QZ survey and Sloan data releases, starting in this investigation with the first release, which overlaps with the 2QZ over an equatorial band.

The 2QZ survey has helped constrain the distribution of DA white dwarfs in the thin Galactic disk with a scale-height of the order of 200-300 pc, but more work remains to be done to improve white dwarf luminosity measurements. Because of large spread in the Sloan colors, the surface gravity, hence mass measurements are uncertain.

With a similar study of non-DA white dwarfs we hope to explore the question of spectral evolution of white dwarf stars, and, in particular, the evolving DA to non-DA population ratio. Moreover, the 6QZ survey based on the UK Schmidt 6dF instrument should help extend the magnitude coverage down to $b_J = 16$ and should help develop a homogeneous sample constraining the local space density of white dwarfs and the scale-height simultaneously.

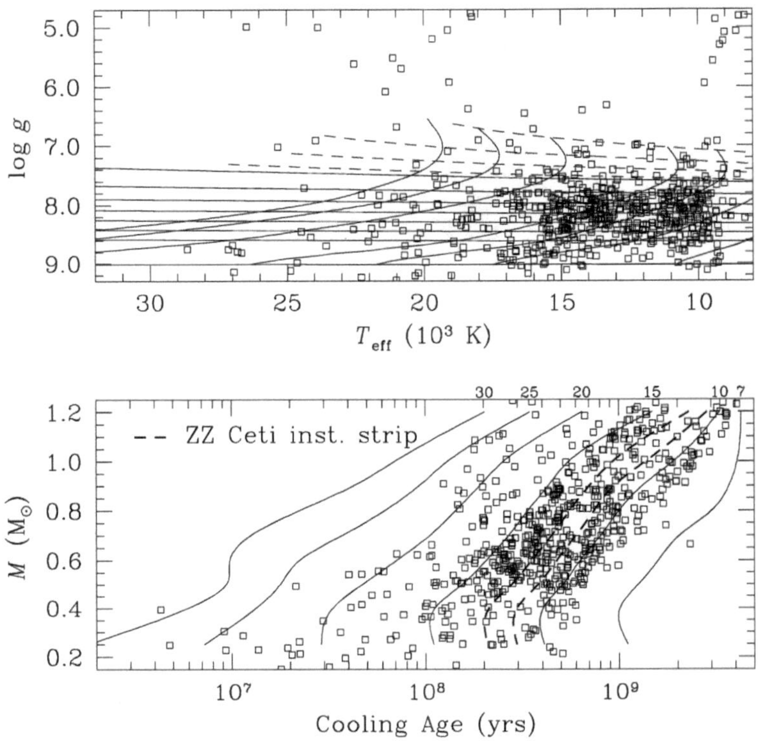

Figure 8. (Top) A surface gravity versus temperature diagram for the joint 2QZ-Sloan sample. The data is compared to evolutionary models from Wood (1995) and Benvenuto & Althaus (1998) between 0.4 and 1.2 $M_\odot$ and 0.25 and 0.4 $M_\odot$, respectively. Cooling ages at 0.3, 0.5, 1.0, 3.0, 5.0, 10.0, and 30×10^8 years, from left to right, are shown crossing the curves of constant masses. (Bottom) A mass versus cooling age diagram for the same sample. Curves of constant temperatures between 30,000 to 7,000 K are shown along with the location of the ZZ Ceti instability strip. Numerous ZZ Ceti candidates are selected.

Acknowledgments

We warmly thank all the present and former staff of the Anglo-Australian Observatory for their work in building and operating the 2dF facility. The 2QZ is based on observations made with the Anglo-Australian Telescope and the UK Schmidt Telescope.

Funding for the creation and distribution of the SDSS Archive has been provided by the Alfred P. Sloan Foundation, the Participating Institutions, the

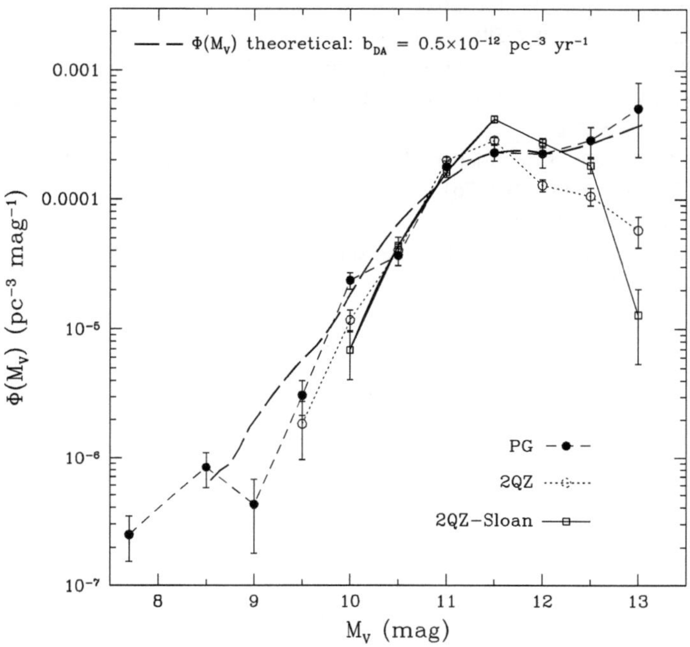

Figure 9. The DA white dwarf luminosity function measured in the Palomar-Green survey (Fleming et al. 1986), the 2QZ survey (Vennes et al. 2002) and the present joint 2QZ-Sloan survey, compared to a model function assuming a constant birthrate. This exercise indicates that the 2QZ-Sloan sample is most useful is defining the luminosity function for $10 \leq M_V \leq 12.5$.

National Aeronautics and Space Administration, the National Science Foundation, the U.S. Department of Energy, the Japanese Monbukagakusho, and the Max Planck Society. The SDSS Web site is http://www.sdss.org/.

References

Benvenuto, O. G. & Althaus, L. G. 1998, *MNRAS*, **293**, 177.

Chen, B., et al. 2001, *ApJ*, **553**, 184.

Croom, S. M., Smith, R. J., Boyle, B. J., Shanks, T., Loaring, N. S., Miller, L., & Lewis, I. J. 2001, *MNRAS*, **322**, L29.

Croom, S. M., Smith, R. J., Boyle, B. J., Shanks, T., Miller, L., Outram, P. J., & Loaring, N. S. 2004, *MNRAS*, **349**, 1397.

Fleming, T. A., Liebert, J., & Green, R. F. 1986, *ApJ*, **308**, 176.

Gilmore, G. & Reid, N. 1983, *MNRAS*, **202**, 1025.

Harris, H. C., et al. 2003, *AJ*, **126**, 1023.

Holberg, J.B., Oswalt, T.D., & Sion, E.M. 2001, *ApJ*, **571**, 512.

Kleinman, S. J., et al. 2004, *ApJ*, **607**, 426.

McCook, G.P., & Sion, E.M. 1999, *ApJS*, **121**, 1.

Raymond, S. N., et al. 2003, *AJ*, **125**, 2621.

Schmidt, G. D., et al. 2003, *ApJ*, **595**, 1101.

Vennes, S. 2003, in White Dwarfs, NATO ASIB Proc. 105 (Dordrecht: Kluwer), 367.

Vennes, S., Smith, R.J., Boyle, B,J., Croom, S.M., Kawka, A., Shanks, T., Miller, L., & Loaring, N. 2002, *MNRAS*, **335**, 673.

Wesemael, F., Greenstein, J. L., Liebert, J., Lamontagne, R., Fontaine, G., Bergeron, P., & Glaspey, J. W. 1993, *PASP*, **105**, 761.

Wood, M.A. 1992, *ApJ*, **386**, 539.

Wood, M. A. 1995, in White Dwarfs, Lecture Notes in Physics 443 (Berlin: Springer Verlag), 41.

WHITE DWARFS AND CATACLYSMIC VARIABLES IN THE FBS

A.M. Mickaelian
Byurakan Astrophysical Observatory (BAO), Byurakan 378433, Aragatzotn province, Armenia.
aregmick@apaven.am

Abstract The Second part of the First Byurakan Survey (FBS) is the continuation of the Markarian Survey and is aimed at discovery of UVX stellar objects: QSOs, Seyferts, white dwarfs, hot subdwarfs, cataclysmic variables, etc. $+33° < \delta < +45°$ and $+61° < \delta < +90°$ regions at $|b| > 15°$ has been covered so far. 1103 blue stellar objects have been selected, including 716 new ones. Observations revealed 17 new white dwarfs and 3 cataclysmic variables, including a new bright (V=12.6) nova-like cataclysmic variable of SW Sex subclass RXS J16437+3402. Highly probable white dwarf candidates have been found having significant proper motions. Polarimetric observations have been undertaken as well: FBS 1704+347 is found to be a possible polar, and FBS 1815+381, a variable magnetic WD. The total number of WDs in the whole survey is estimated to be 270 (24%), and cataclysmic variables – 45 (4%).

Keywords: Surveys, white dwarfs, cataclysmic variables

1. The First Byurakan Survey - *FBS*

The First Byurakan Survey have been carried out by B.E. Markarian, V.A. Lipovetski and J.A. Stepanian in 1965-1980 with the Byurakan Observatory 1m Schmidt telescope using 1.5° prism (Markarian et al. 1989). 2050 Kodak IIaF photographic plates in 1133 fields ($4° \times 4°$ each) have been taken. FBS covers 17,000 deg^2 area with $\delta > -15°$ and $|b| > 15°$. The limiting magnitude is 17.5^m-18^m. The scale is 96.8"/mm and the dispersion is 1800 Å/mm near Hγ and 2500 Å/mm near Hβ (mean spectral resolution being about 50 Å). Low-dispersion spectra cover the range 3400-6900 Å, and there is a sensitivity gap near 5300 Å, dividing the spectra into red and blue parts. It is possible to compare the red and blue parts of the spectrum (easily separating red and blue objects), follow the spectral energy distribution, notice some broad emission and absorption lines, thus making up some understanding about the nature of the objects. Each FBS plate contains low-dispersion spectra of some 15,000-20,000 objects, and there are some 20,000,000 objects in the whole survey.

E.M. Sion, S. Vennes and H.L. Shipman (eds.), White Dwarfs: Cosmological and Galactic Probes, 61-72.
© 2005 *Springer*.

The FBS was conducted originally for search for galaxies with UV-excess. The discovery of 1515 UV-excess (UVX) galaxies by Markarian and colleagues (later called Markarian galaxies) was the first and the most important work based on the FBS plates (Markarian et al. 1989). The study of Markarian galaxies brought to discovery of many new Seyferts and the first spectral classification of this type of objects, as well as to the definition of starburst galaxies. There are more than 200 AGN, radio, IR, X- and gamma-ray sources, interacting and merging objects, galaxies with double and multiple nuclei among the Markarian galaxies. The Markarian survey was the first systematic survey for AGNs, it was a new method of search for AGNs. The catalog of Markarian galaxies is available at CDS.

2. The *FBS* blue stellar objects

The second part of the FBS was devoted to discovery and study of blue (or UV-excess) stellar objects (Abrahamian & Mickaelian 1996; Mickaelian 2000). It was carried out in 278 FBS fields, in a 4009 deg^2 area of the FBS (+33°<δ<+45° and δ>+61°). The main purpose of this work was discovery of new bright QSOs, Seyferts, planetary nebulae nuclei, cataclysmic variables (CV), white dwarfs (WD), subdwarfs, etc. This survey is similar to the Palomar-Green (PG) one (Green et al. 1986), however, the method of selection is spectroscopic, which is more efficient. On the other hand, FBS goes deeper in the limiting magnitude and covers a larger area.

Markarian's criteria for classification of low-dispersion spectra concerned the extended objects. We have revised slightly these criteria and applied for the blue stellar objects (BSOs). The presence of the green gap in objective-prism spectra allows select and classify the spectra by the intensity and length correlation of the two divided parts (conditionally, "blue" and "red"). The B type (blue) designates objects having the blue part of the spectrum more intense than the red one, and the N type (neutral) - objects having approximately equal intensities of both blue and red parts. The indices 1, 2 and 3 show the ratio of the lengths of the blue and the red parts in decreasing order. The indices "a" and "e" show the presence of absorption and emission lines, respectively. Index "v" stands for variability. A colon denotes the cases of uncertainty of these data. Thus, the bluest objects with the strongest UV-excess are classified as B1. In all, FBS lists contain 396 B1-type objects, 342 B2-type, 124 B3, 105 N1, 90 N2, and 38 N3. A small number (8) of extended objects having classification like Markarian galaxies (s, sd, ds, and d) have been included as well. 190 have absorption features, and 104, emission features. 7 show evidence of variability.

By these criteria we find early-type main sequence stars (O-B5), horizontal-branch stars (HBB), hot subdwarfs of sdO and sdB classes, white dwarfs of

DO, DB, DA, DC and DZ classes, planetary nebulae nuclei (PNN), cataclysmic variables (CV), binaries (Bin), as well as extragalactic objects: QSO candidates, Seyfert galaxies with star-like image, BL Lac objects, compact galaxies (BCDGs, etc.), other emission-line and UV-excess galaxies and other peculiar objects. In all, 1103 objects have been selected, including 716 new blue stellar objects. The FBS catalog is available at CDS. It contains also estimation of the class for some part of objects: DA, DB, PNN, CV, Bin, QSO, Sy, Gal; in all for 202 objects.

We have cross-correlated our sample with the MAPS catalog (Cabanela et al. 2003) and found 913 associations (190 objects have $|b|<20°$ and not appear in MAPS). The APS color-magnitude diagram for these 913 FBS objects is given in Fig. 1.

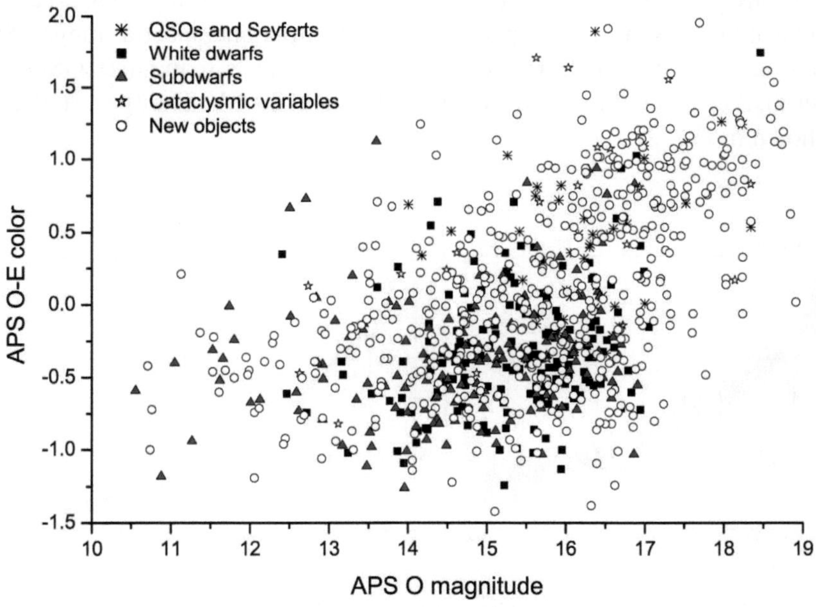

Figure 1. APS color-magnitude diagram for 913 FBS blue stellar objects.

3. Spectroscopic observations

Though more than 400 FBS blue stellar objects have been observed, only a small number of them have been published so far (Abrahamian & Mickaelian 1991; Mickaelian et al. 1999; 2001; 2002).

Spectroscopic follow-up observations of some part of the sample have been made mainly with the BAO 2.6m telescope during 1987-1991. The UAGS spectrograph attached to the Cassegrain focus has been used. The linear dispersion was 101 Å/mm and the spectral range, 3350-6050 Å. For the classification, we have adopted criteria given in Sion et al. (1983) and used by Green et al. (1986), to compare our survey with PG. Out of 54 classified objects, 11 are white dwarfs (4 DA, 1 DBA, 1 DB, 2 DAZ, 2 DZ, 1 DC). A number of observations have not been reduced yet. In 1996, 5 FBS objects were observed at the Russian SAO 6 m telescope with the UAGS spectrograph and TK 530×580 CCD. These results have not yet been published.

Observations with the Observatoire de Haute-Provence (OHP) 1.93m telescope have been made in 1997-1998 for search for new QSOs and Seyferts. The spectrograph CARELEC (Lemaitre et al. 1989) with a TK 512×512 CCD have been used. The dispersion was 7.1 Å/pix and spectral range 3800-7350 Åhas been obtained. Out of 37 observed objects (candidate QSOs and Seyferts), 3 turned out to be DA WDs, 2 - DZ, and 2 - CV. The spectra of 2 DA are shown in Fig. 2.

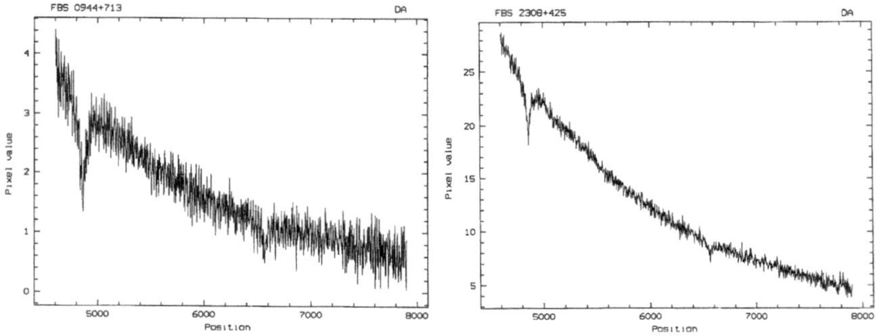

Figure 2. Spectra of 2 FBS white dwarfs, obtained with the OHP 1.93m telescope.

New spectroscopic CCD observations have been carried out in Byurakan with the 2.6m telescope since 1998. The long-slit spectral camera ByuFOSC with the "green" grism (600 mm^{-1}) and TM 1060×1028 CCD has been used. The dispersion was 2.7 Å/pix, the spectral range – 4200-6900 Å. The program was aimed at detection of new QSOs and Seyferts, as well as new WDs and CVs. 7 objects have been observed, 1 being a DA WD.

The program of search for FBS AGN included a cross-correlation of radio and X-ray (ROSAT, Voges et al. 1999; 2000) sources with the USNO (Monet et al. 1996) objects and their further inspection in the FBS plates. A number of additional blue stellar objects have been found and observed spectroscopically in Byurakan. One of them, RXS J16437+3402, turned to be a wonderful bright (V=12.6) nova-like cataclysmic variable of SW Sex subclass, having a period within the period "gap" for such objects. It was discovered in Byurakan in 2000, and then studied in detail photometrically and spectroscopically at OHP (Mickaelian et al. 2002). Fig. 3 gives its normalized spectrum, where strong Hβ and Hγ lines, the Bowen blend (at λ4645Å), HeII λ4686Å, and a number of HeI lines are seen, most of them having a strong narrow emission component inside the broad absorption one.

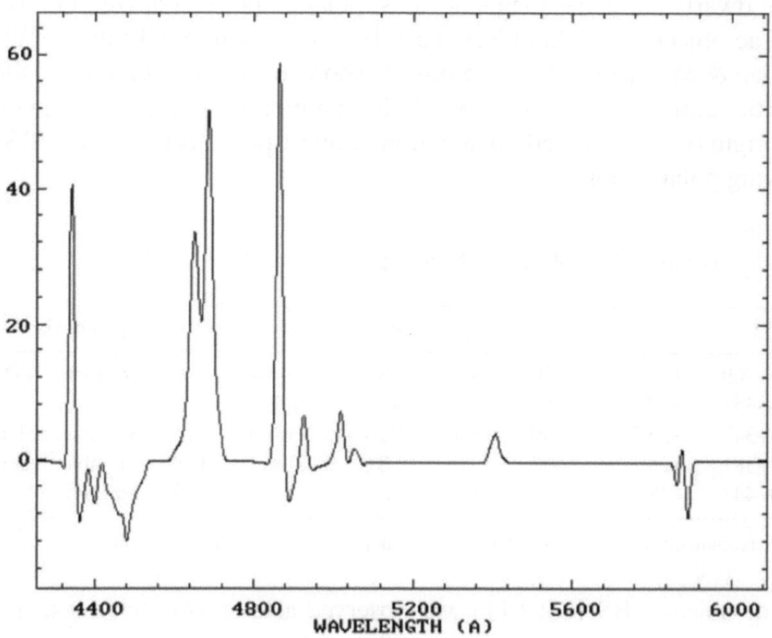

Figure 3. The normalized spectrum of the cataclysmic variable RXS J16437+3402.

Only 17 FBS white dwarfs and 3 CVs have been published so far. However, new lists will be published after the full reduction of the obtained data. Preliminarily, ~100 new WDs and 7 CVs are among the observed objects. We can estimate the fraction of different types in FBS sample based on known objects (both associated with objects from available catalogs and obtained from our spectroscopic observations), as well as our preliminary types marked in the catalog. The distribution is as follows:

hot subdwarfs (sdB, sdB-O, sdOA, sdOB, sdOC, sdOD, sdO)	48 %
white dwarfs (DA, DB, DBA, DZ, DAZ, DBZ, DC, DO, PNN)	24 %
main sequence O-B, and HBB stars	8 %
cataclysmic variables and other binaries	4 %
extragalactic objects (QSO, Sy, galaxies)	8 %
objects with continuous spectra (possibly DC, BLL, QSO)	3 %
not classified (low quality spectra)	5 %

4. Polarimetric observations

Polarimetric observations have been carried out in 1990-91 with the Byurakan 2.6m telescope. This study is aimed at revealing the intrinsic polarization of new objects and discovery of new objects of definite types: magnetic white dwarfs and other magnetic stars, polars, highly polarized QSOs (HPQ), BL Lac objects, etc. 35 objects have been observed (Mickaelian et al. 1991; Eritsian & Mickaelian 1993). 5 objects show linear polarization and their high galactic latitudes and relative proximity mention on intrinsic (or environmental) origin of the observed polarization. Table 1 presents the list of FBS objects showing polarization.

Table 1. The list of FBS objects with linear polarization.

FBS	*b*	r_{max}*	*B*	*Filters*	*P(%)*	*θ(°)*	*Comments*
1559+369	+49.1	70	14.2	B	3.5-3.9	180	magnetic WD
1654+351	+37.8	350	12.7	B, U	0.7	85	
1704+347	+35.7	1200	15.4	B, 0	0.5-6.0	16	variable, polar?
1815+381	+22.7	400	13.0	B	2.5-4.4	178	variable, magnetic WD
1850+443	+18.4	160	11.0	B	2.0-3.3	135	variable

* the maximum distance is estimated based on the upper limit of absolute magnitude

One object, FBS 1850+443 was observed also at SAO 6m telescope with a hydrogen magnetometer (Fabrika 1993) and the presence of the linear polarization has been confirmed.

5. The sample of *FBS* white dwarfs

White dwarfs are rather easy to notice by the low-dispersion spectral features, as their broad absorption lines make them in fact the only objects showing Balmer series (Hβ, Hγ and Hδ are especially well seen), as narrow lines are lost due to low spectral resolution. However, in a number of cases we are not able to detect these lines because of the low contrast. WDs, together with the hot subdwarfs, are the bluest objects, and the comparison of red and blue parts of the FBS spectra distinguishes them among other types. Finally, WDs show often proper motions (PM) (if their space motion is oriented appropriately), and this is an additional criterion for their detection.

To verify our selection and detection of different types of objects, we have analyzed FBS spectra for known types. On the basis of spectral classes of the 387 re-discovered known objects from catalogs and other surveys, as well as our follow-up spectral observations, we have compared the spectral classes with the preliminary types of low-dispersion spectra. White dwarfs make up 77% of B objects with absorption features and 34% of B objects without absorption features (as mentioned, absorption lines are not always noticed). So we should search white dwarfs among the B type, especially with absorption lines (B1a, B2a, B3a). In all, we have re-discovered 141 known white dwarfs from the Catalog of Spectroscopically Identified White Dwarfs (hereafter, MCS) (McCook & Sion 2003).

We have carried out accurate astrometric measurements for the whole FBS sample (1103 objects) using the DSS1 and DSS2 (both red and blue) images. This method has been proved to give 0.6" accuracy in both α and δ (Véron-Cetty & Véron 1996). A comparison of DSS1 and DSS2 measurements revealed a large number of objects with positional differences between the two epochs. To be secure, we have accepted as a proper motions (PM) only those cases, when we have a difference >3" (however, most of >2" differences, in all 231 cases, should be due to PM as well). Most of these objects show differences in configuration so that it is noticed by eye (Fig. 4). In one case, FBS 574, we were able to detect a large difference even between DSS2r and DSS2b (a positional change of 3.88" during 5.9 years). However, this depends on the separation between the DSS2r and DSS2b observations, which is not always large. In case of DSS1 and DSS2, the separation is 30 years and larger (33-47 for our sample).

We have checked the DSS1 and DSS2 images of objects showing >3" positional differences with a large zoom having the object in the exact center to select those with obvious PM. Out of 105, we have selected 77 such objects with confirmed shift of their positions. Others may be due to different plate solution of DSS1 and DSS2, or other effects. We have checked these 77 objects in SIMBAD: 57 are known WDs and only 1 is a subdwarf, 19 being unknown. We list these 19 objects, highly probably WDs, in Table 2.

Out of 57 known WDs, only 16 had measured PM in the WD catalog, and we have obtained for the first time PM for 41 others too. Considering the DSS1 accuracy as 0.6", and the DSS2r, 0.4" (according to our estimate), we can roughly accept 1" difference as the uncertainty upper limit. Thus we should accept 1" difference divided by the separation in years as the accuracy of PM

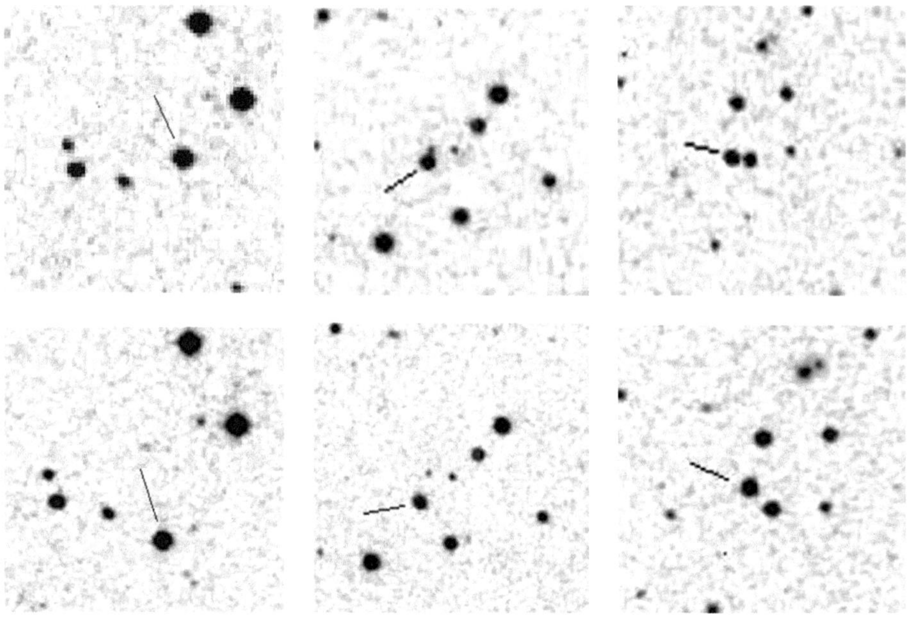

FBS 574 FBS 654 FBS 1033

Figure 4. DSS1 and DSS2r images of 3 FBS objects showing large proper motion.

measurement. It is 0.021-0.030 "/yr for our objects. In Fig. 5, we compare our data with previously known ones for 16 WDs having known PM.

In the Catalog of WDs (McCook & Sion 2003), distances are known for 289 objects, including 244 (84%) within 50 pc and 273 (94%) within 100 pc. PM are known for 1293 objects, including 272 having distance measurements. For all these WDs we have calculated the tangential velocities: 227 (83%) have velocities within 100 km/sec, and 258 (95%), within 200 km/sec). Accepting 100 km/sec as an upper limit for v_t, we obtain that an object having PM=0.1 (the typical value for our list of 19 objects), should be at a distance of less than 210 pc and for O=15^m have M_{abs} fainter than 8.4^m. Given that the objects are blue, this makes them either WDs or subdwarfs. However, the v_t limit is larger than the typical values (v_t~50 km/sec), so we consider these objects as highly probable WDs.

Thus subsample of white dwarf candidates can be made up on the basis of the low-dispersion spectral types, and PM. The main idea is to select candidates for further purposeful investigations before the spectral observations of all FBS objects. ~270 WDs are estimated to be in the FBS sample. The subsample of

Table 2. The list of 19 FBS stars having large PM, new candidate WDs.

FBS #	*FBS* name	*APS* O	*APS* O-E	*Type*	*DSS2-DSS1* difference	*PM* "/yr	*PA* °	M_{abs}* >
449	0106+353	15.0:		B2	9.5	0.239	337	10.3
20	0150+396	16.0:		B2a:	4.4	0.122	132	9.8
460	0213+355	14.77	-0.72	B2	3.8	0.109	205	8.3
654	0712+623	15.11	-0.20	B2	6.3	0.143	173	9.3
660	0742+625	11.61	-0.60	B1	4.1	0.094	8	4.9
872	0817+721	15.55	-0.61	B2a	6.9	0.163	191	10.0
885	1048+715	15.89	-0.39	B3a	6.4	0.140	224	10.0
688	1103+619	15.9:	-0.1:	B3	4.4	0.116	170	9.6
811	1231+680	17.04	-0.29	N2	5.0	0.110	291	10.6
711	1322+627	16.06	-0.64	B2	3.5	0.078	167	8.9
1012	1344+765	16.3:	-0.2:	B2	4.2	0.101	240	9.7
109	1403+386	15.86	-0.15	B2	4.7	0.118	170	9.6
1015	1405+749	15.33	-0.45	B1a	3.6	0.088	220	8.4
746	1645+649	15.84	-0.32	B1a:	3.4	0.089	181	9.0
158	1749+393	16.01	-0.87	B1a:	3.1	0.078	220	8.9
613	2154+329	13.7:	-0.3:	B1	3.4	0.090	87	6.9
620	2212+335	15.3:	-0.4:	B1	5.3	0.159	146	9.7
189	2254+373	16.44	-0.04	B2a	5.1	0.148	119	10.7
203	2354+375	16.38	0.14	B1	3.2	0.080	150	9.3

* the estimated upper limit for M_{abs} for v_t=100 km/sec.

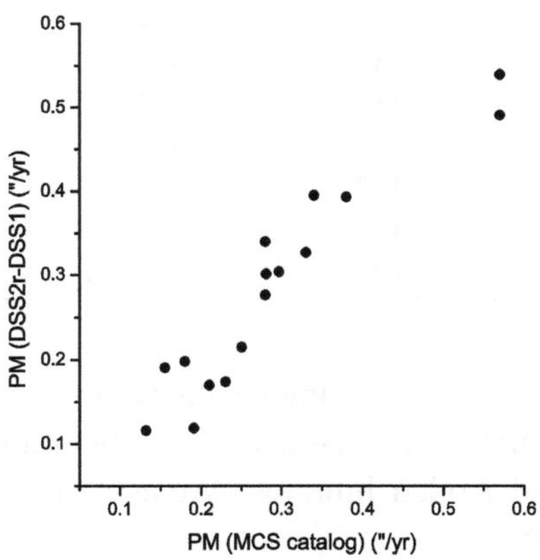

Figure 5. A comparison of previously known and newly measured PM for 16 WDs.

WDs may reveal new pulsating WDs (ZZ Ceti stars), magnetic WDs, AM Her type objects (polars), PNN (DO, PG 1159 type), etc.

6. The sample of *FBS* cataclysmic variables

Cataclysmic variables (CVs) are rather difficult to notice by the low-dispersion spectral features, as their emission-lines are narrow and often lost due to low spectral resolution. However, in a number of cases we were able to detect these lines even during the eye selection of objects. CVs are also very blue, and often show variability comparing the FBS different plates or FBS with the DSS (as well as DSS1 and DSS2).

A subsample of the FBS candidate CVs has been compiled using the mentioned criteria. The list of these objects is given in Table 3.

Table 3. The list of 17 FBS objects, candidate CVs.

FBS #	*FBS* name	*FBS* B	*APS* O	*APS* O-E	*FBS* type
437	0039+361	15.7	15.63	-0.50	B2e
218	0104+424	16.0			B1v
468	0306+333	14.7			B1
473	0335+344	15.6			N1e
1048	0657+888	15.0	16.03	1.64	N3e
655	0713+648	16.0	16.53	1.92	N1v:
483	0742+339	15.9	14.64	-0.45	B1
974	0827+738	15.9	16.93	0.81	N3e
100	1327+411	15.5			N2v
1075	1328+881	16.2	16.56	1.08	N3e
1076	1330+837	14.9	16.40	1.09	N3e
1084	1441+827	15.6	16.15	0.82	N2e:
906	1614+711	16.4	16.72	-0.14	B2e:
156	1738+372	14.5	16.13	-0.39	B2v:
386	1745+420	13.5	17.32	1.22	B2v:
632	2318+341	15.3	15.01	-0.29	B2
428	2342+432	14.0			B1

In all, FBS has re-discovered 21 known cataclysmic variables from the Catalogue of cataclysmic variables, low-mass X-ray binaries and related objects (Ritter & Kolb 2003) and 7 other binaries. We estimate that in total there are ~25 new CVs in the FBS.

It is worth mentioning that we have found 2 emission-line blue objects during the inspection of the FBS plates, which are absent in both DSS1 and DSS2 (both red and blue) charts. One has $\sim 13.5^m$ FBS B magnitude, and the other, $\sim 14^m$. Given that DSS2b goes up to 22^m, the estimated lower limit for the amplitude is 8.5^m and 8^m, respectively. We believe these are Novae, and plan to make a deep imagery in these fields to detect the progenitor stars.

7. *DFBS* and future studies

The project of the Digitized First Byurakan Survey (DFBS) is active currently and will be finished in 2004 (Mickaelian et al. 2003). The DFBS catalog will contain data on some 20,000,000 objects brighter than 17.5^m. It will give possibility for search for new bright QSOs and Markarian (UVX) galaxies, new blue stellar objects (continuation of the 2nd part of the FBS), new BCDGs, optical counterparts of IR, radio, and X-ray sources (optical identifications), late-type stars, planetary nebulae, emission line stars (especially CVs), and others. FBS is particularly useful as a source for search for white dwarfs. Their broad absorption lines make them easily detectable among the other objects. Beside the continuation of the 2nd part of the FBS, where significant number of WDs is expected, non-UVX white dwarfs will be selected by their absorption lines only. Numerous such objects have been noticed during the inspection of plates, however not included in the FBS, as they are not blue.

We have started also optical identifications of ROSAT X-ray sources (Voges et al. 1999; 2000) with the digitized Hamburg Quasar Survey (Engels et al. 2001). A number of new WDs and CVs are present among the blue objects matching ROSAT sources. Identifications of X-ray sources using the DFBS are planned as well, and new WDs and CVs are expected.

A combination of all selection methods (selection of UV-excess, absorption and emission-line objects, objects having blue color in MAPS, optical identification of X-ray sources) using the DFBS will bring to a complete sample of high galactic latitudes WDs and CVs up to 17.5^m.

References

Abrahamian, H.V., Mickaelian, A.M. 1991, *Afz*, **35**, 197.
Abrahamian, H.V., & Mickaelian, A.M. 1996, *Ap*, **39**, 531.
Cabanela, J.E., Humphreys, R.M., Aldering, G., et al. 2003, *PASP*, **115**, 837.
Engels, D. et al. 2001, *ASP Conf. Series*, **232**, 326.
Eritsian, M.H., Mickaelian A.M. 1993, *Afz*, **36**, 203.
Fabrika, S.N. 1993, private communication.
Green, R.F., Schmidt, M., & Liebert, J. 1986, *ApJS*, **61**, 305.
Lemaitre, G., Kohler, D., Lacroix, D., Meunier, J.-P., Vin, A. 1989, *A&A*, **228**, 546.
Markarian, B.E., Lipovetski, V.A., Stepanian, J.A., Erastova, L.K., Shapovalova, A.I. 1989, *Commun. Special Astrophys. Obs.*, **62**, 5.
McCook, G.P., Sion, E.M. 1999, *ApJS*, **121**, 1.
Mickaelian, A.M. 2000, *AATr*, **18**, 557.
Mickaelian, A.M, Eritsian, M.H, Abrahamian, H.V. 1991, *Afz*, **34**, 351.
Mickaelian, A.M, Gonçalves, A.C., Véron-Cetty, M.P., Véron P. 1999, *Ap*, **42**, 1.
Mickaelian, A.M, Gonçalves, A.C., Véron-Cetty, M.P., Véron P. 2001, *Ap*, **44**, 14.

Mickaelian, A.M., Balayan, S.K., Ilovaisky, S.A., Chevalier, C., Véron-Cetty, M.P., Véron, P. 2002, *A&A*, **381**, 894.

Mickaelian, A.M., et al. 2003, IAU Symp. 216: Maps of the Cosmos, ASP (in press).

Monet, D., Bird, A., Canzian, B., et al. 1996, USNO-A2.0, Washington DC.

Ritter, H. & Kolb, U. 2003, *A&A*, **404**, 301.

Sion, E.M., Greenstein, J.L., Landstreet, J.D., Liebert, J., Shipman, H.L., Wegner, G.A. 1983, *ApJ*, **269**, 253.

Véron-Cetty, M.P. & Véron, P. 1996, *A&A*, **115**, 97.

Voges, W., Aschenbach, B., Boller, T., et al. 1999, *A&A*, **349**, 389.

Voges, W., Aschenbach, B., Boller, T., et al. 2000, MPE, Garching.

SPECTROSCOPIC AND PHOTOMETRIC OBSERVATIONS OF WHITE DWARFS IN GLOBULAR CLUSTERS

S. Moehler[1,2], D. Koester[1], U. Heber[2], R. Napiwotzki[2], M. Zoccali[3], F.R. Ferraro[4], A. Renzini[3]

[1]*Institut für Theoretische Physik und Astrophysik, Universität Kiel, 24098 Kiel, Germany*

[2]*Dr. Remeis-Sternwarte, Astronomisches Institut der Universität Erlangen-Nürnberg, Sternwartstr. 7, 96049 Bamberg, Germany*

[3]*European Southern Observatory, Karl Schwarzschild Strasse 2, 85748 Garching bei München, Germany*

[4]*Dipartimento di Astronomia, Università di Bologna, via Ranzani 1, 40127 Bologna, Italy*

Abstract White dwarfs in globular clusters offer additional possibilities to determine distances and ages of globular clusters, provided their spectral types and masses are known. We therefore started a project to obtain spectra of white dwarfs in the globular clusters NGC 6397 and NGC 6752. All observed white dwarfs show hydrogen-rich spectra and are therefore classified as DA. Analysing the multi-colour photometry of the white dwarfs in NGC 6752 yields an average gravity of log g = 7.84 and 0.53 $M_\odot$ as the most probable average mass for globular cluster white dwarfs. Using this average gravity we try to determine independent temperatures by fitting the white dwarf spectra. While the stellar parameters determined from spectroscopy and photometry usually agree within the mutual error bars, the low resolution and S/N of the spectra prevent us from setting constraints stronger than the ones derived from the photometry alone, apart from verifying the hydrogen-rich nature of the targets.

Keywords: Globular clusters, Galactic halo

1. Introduction

White dwarf sequences in globular clusters can be used as standard candles for determining the distance, similarly to the traditional main sequence fitting procedure (Renzini & Fusi Pecci 1988, Renzini et al. 1996). Despite their faintness white dwarfs offer some advantages as standard candles when compared to main sequence stars:

E.M. Sion, S. Vennes and H.L. Shipman (eds.), White Dwarfs: Cosmological and Galactic Probes, 73-77.
© 2005 *Springer.*

- They come in just two varieties - either hydrogen-rich (DA) or helium-rich (DB) – *independent of their original metallicity* and, in both cases, their atmospheres are virtually free of metals. So, unlike in the case of main sequence fitting, there is not the problem of finding local calibrators with the same metallicity as the globular clusters.
- White dwarfs are locally much more numerous than metal-poor main sequence stars and thus allow to define a better reference sample.

However, the method has its own specific problems, as the location of the white dwarf cooling sequence depends on the mass, the envelope mass, and the spectral type of the white dwarfs (see Zoccali et al. 2001 and Salaris et al. 2001 for more details). We therefore observed spectra of white dwarfs in NGC 6397 and NGC 6752 in order to determine their spectral types and get mass estimates.

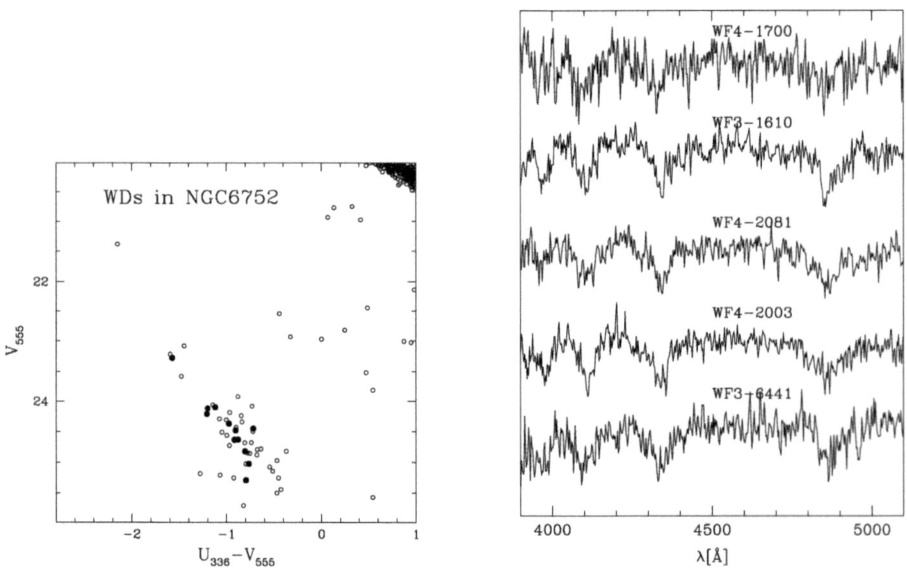

Figure 1. Colour-magnitude diagram and normalized spectra of the white dwarfs in NGC 6752. The filled circles mark the twelve most isolated white dwarfs, whose photometry has been analysed and from which the spectroscopic targets were selected

2. Observations and Data Reduction

The photometric dataset used in the present work consist of a series of HST/WFPC2 exposures taken through the filters F336W, F439W, F555W, F814W as a part of the HST programme GO5439. Results based on this dataset have been already published in Renzini et al. (1996) and Ferraro et al. (1997) and the data reduction is discussed by Moehler et al. (2004). The spectroscopic targets in NGC 6397 were selected from the photometry by Cool et al. (1996), those in NGC 6752 from the data mentioned above. We took care to select stars as isolated as possible in order to avoid contamination from neighbouring stars. The spectra were obtained in service mode with FORS1 at VLT-UT1 (Antu) and VLT-UT3 (Melipal) and their reduction is described in detail in Moehler et al. (2004).

3. Fit of the photometry of NGC 6752

As we know from the spectra that all observed white dwarfs have hydrogen-rich atmospheres we can use the multi-colour photometry to determine $T_{\rm eff}$ and log g for the stars in NGC 6752. To calculate theoretical magnitudes in the so called "STMAG" system of the WFPC2 (Holtzman et al. 1995) we used the grid of DA model atmospheres of Koester (see Finley et al. 1997 for a detailed description) and the following definition of $M_{\rm STMAG}$

$$M_{\rm STMAG} = -2.5\,\frac{\int S(\lambda)F_\lambda\,d\lambda}{\int S(\lambda)d\lambda} - 21.1$$

where $S(\lambda)$ is the filter response function and F_λ is the mean intensity of the stellar disk. Note that $M_{\rm STMAG}$ is not an absolute magnitude but used to distinguish theoretical from observed magnitudes m.

When trying to fit a set of observed magnitudes with theoretical values we have in principle the possibility to determine simultaneously the three parameters $T_{\rm eff}$, log g and distance. Due to observational errors and the strong degeneracy with respect to the the parameters, it is impossible to determine all three free parameters from only four observed magnitudes. The *relative* energy distribution for the white dwarfs in the observed parameter range depends mostly on $T_{\rm eff}$ and only very little on log g. The *solid angle* depends strongly on both log g (through the mass-radius relation) and on the distance. Therefore a small change in log g can be easily compensated by a change in distance and vice versa.

We therefore decided to keep the distance modulus fixed as it is rather well known for NGC 6752. Before the fit, the magnitudes were dereddened assuming E_{B-V} = 0.04, and the extinction coefficients of Holtzman et al. (1995). The average surface gravity is 7.96 for $(m\text{-}M)_0$ = 13.05 and 7.84 for $(m\text{-}M)_0$ = 13.20 with an error of the mean of 0.02. With the Wood (1995) mass-radius relation for a typical $T_{\rm eff}$ of 15000 K and "thick hydrogen layer" (10^{-4} of the stellar mass) this corresponds to masses of 0.59 $M_\odot$ and 0.53 $M_\odot$, for the short and long distance modulus, respectively. Assuming instead the more unlikely case of a "thin" hydrogen layer ($\leq 10^{-6}$ $M_{\rm WD}$) the mass would be 0.56 $M_\odot$ for $(m\text{-}M)_0$ = 13.05 and 0.50 $M_\odot$ for $(m\text{-}M)_0$ = 13.20. If we consider that the progenitor stars in globular clusters should have less than 1 $M_\odot$ and that NGC 6752 has an exclusively blue horizontal branch with a very extended

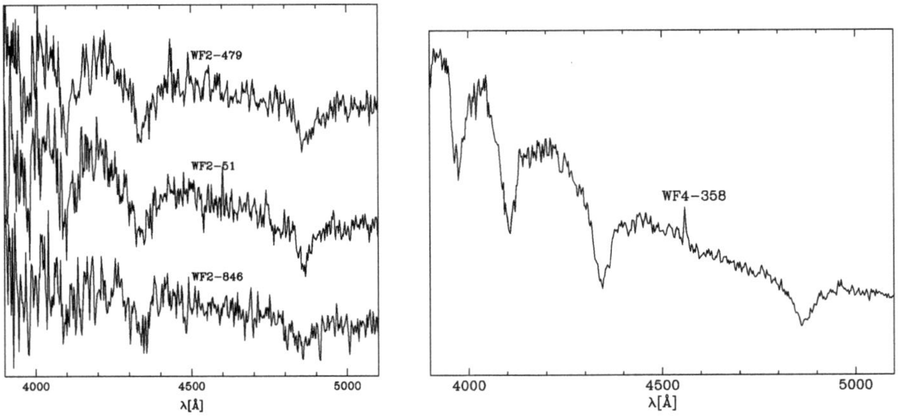

Figure 2. Spectra of the white dwarfs in NGC 6397 (WF4-358 is brighter than the other ones by more than 1 magnitude)

blue tail, the smaller mass seems much more likely and gives strong support for the larger distance.

4. Spectral fits for NGC 6752 and NGC 6397

The observed spectra for the white dwarfs in the two clusters were analysed with the same model atmosphere grid used for the calculation of the theoretical magnitudes. The low S/N and resolution of the spectra leads to very large errors if T_{eff} and $\log g$ are both used as free parameters. We have therefore decided to keep $\log g$ fixed at the value 7.84 obtained as the most likely value from the photometry of NGC 6752. The effective temperatures derived for the white dwarfs in NGC 6752 usually agree with the photometric temperatures within the mutual error bars. For the stars in NGC 6397 a comparison between spectroscopic and photometric parameters is not possible as only V and I photometry exists. The spectra of the three fainter white dwarfs have too low S/N to permit a fit of both T_{eff} and $\log g$. The spectroscopic analysis of the brightest white dwarf WF4-358 yields parameters that are inconsistent with its cluster membership and it may therefore be a field star.

5. Conclusions

Spectroscopy of white dwarfs in the globular clusters NGC 6397 and NGC 6752 showed that all targets are hydrogen-rich DA. From multicolour photometry we determined an average mass of $0.53\,M_{\odot}$, which is identical to that assumed by Renzini et al. (1996) on theoretical grounds. Therefore both observational and theoretical arguments strongly advocate against the use of 0.6 $M_\odot$ (the mean mass of the local white dwarfs) or even more for the mass of white dwarfs in globular clusters when comparing observations to theoretical tracks.

Acknowledgments

We want to thank the staff at Paranal observatory for their great effort in performing these demanding observations. While in Bamberg S.M. was supported by the BMBF grant No. 50 OR 9602 9.

References

Cool, A.M., Piotto, G., King, I.R. 1996, *ApJ*, **468**, 655.

Ferraro, F.R., Carretta, E., Bragaglia, A., Ortolani, S., Renzini, A. 1997, *MNRAS*, **286**, 1012.

Finley D.S., Koester D., Basri G. 1997, *ApJ*, **488**, 375.

Holtzman, J.A., Burrows, C.J., Casertano, S., Hester, J.J., Trauger, J.T., Watson, A.M., Worthey, G. 1995, *PASP*, **107**, 1065.

Moehler, S., Koester, D., Zoccali, M., Ferraro, F.R., Heber, U., Napiwotzki, R., Renzini, A. 2004, *A&A*, **420**, 515.

Renzini, A., Fusi Pecci, F. 1988, *ARA&A*, **26**, 199.

Renzini, A., Bragaglia, A., Ferraro, F.R., et al. 1996, *ApJ*, **465**, L23.

Salaris, M., Cassisi, S., García-Berro, E., Isern, J., Torres, S. 2001, *A&A*, **371**, 921.

Wood, M.A. 1995, Theoretical White Dwarf Luminosity Functions: DA Models, in White Dwarfs, eds. D. Koester & K. Werner, Lecture Notes in Physics, (Springer-Verlag, Berlin Heidelberg New York), Vol. 443, p.41.

Zoccali, M., Renzini, A., Ortolani, S., et al. 2001, *ApJ*, **553**, 733.

THE CORE/ENVELOPE SYMMETRY IN PULSATING WHITE DWARF STARS

T. S. Metcalfe[1], M. H. Montgomery[2], D. E. Winget[3]
[1]*Harvard-Smithsonian Center for Astrophysics*
[2]*Institute of Astronomy, University of Cambridge*
[3]*Department of Astronomy, University of Texas-Austin*

Abstract Recent attempts to model the variable DB white dwarf GD 358 have led to the discovery that two distinct physical descriptions of the interior structure can both match the observed pulsation periods at roughly the same level of precision. One model, inspired by time-dependent diffusion calculations, contains two composition transition zones in the envelope. The other, motivated by the nuclear burning history of the progenitor, includes one transition zone in the core and another one in the envelope. We demonstrate that the non-uniqueness of our modeling framework for these stars originates from an inherent symmetry in the problem. We provide an intuitive example of the source of this symmetry by analogy with a vibrating string. We establish the practical consequences both analytically and numerically, and we verify the effect empirically by cross-fitting two structural models. Finally, we show how the signatures of composition transition zones brought about by physically distinct processes may be used to help alleviate this potential ambiguity in our asteroseismological interpretation of the pulsation frequencies.

Keywords: stellar evolution, interiors, oscillations, white dwarfs

1. ONE STAR, TWO MODELS

Recently, two alternative structural models have been proposed for the most thoroughly observed helium-atmosphere variable (DBV) white dwarf GD 358. Both models reproduce the observations with a comparable level of precision (see Figure 1). There is good reason to believe that the physical basis of each model is sound, but that neither represents a complete description of the true interior structure.

2. SIMPLE ANALOGY: VIBRATING STRING

How can we understand the ability of two physically distinct models to produce reasonable fits to the periods observed in GD 358? Begin by consid-

E.M. Sion, S. Vennes and H.L. Shipman (eds.), White Dwarfs: Cosmological and Galactic Probes, 79-83.
© 2005 *Springer*.

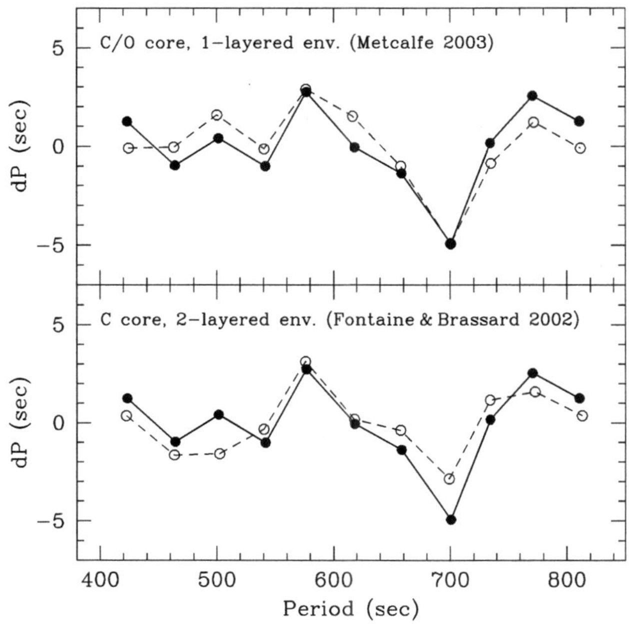

Figure 1. The pulsation periods observed in GD 358 (solid points) plotted against their deviations from the mean period spacing (dP), along with two physically distinct model fits (open points). The fit of Metcalfe (2003) has extra structure in the core (top panel), while the fit of Fontaine & Brassard (2002) has extra structure in the envelope (bottom panel).

ering the oscillations of a uniform-density vibrating string, which produces a spectrum of evenly spaced eigenfrequencies. If we introduce a small density perturbation at some distance (x) from one end of the string, it will change the spectrum of eigenfrequencies slightly. The same perturbation introduced at the same distance from the opposite end of the string ($L-x$) will yield an identical set of perturbed eigenfrequencies. If the density of the string is instead a linear function of position along the string, then the symmetry axis will move toward one end of the string; but it will still be possible to make a (slightly more complicated) density perturbation on the other side of the symmetry axis to produce an identical set of perturbed eigenfrequencies (Montgomery et al. 2003).

3. DBV CORE/ENVELOPE SYMMETRY

In this context, a white dwarf is like an *extremely* non-uniform vibrating string, with the symmetry axis far out in the envelope. This inherent symmetry leads to an ambiguity in the location of internal structures that produce mode trapping in DBV white dwarf stars. In Figure 3 we see that structure in the core near a fractional mass of $\sim$0.5 can produce mode trapping that is similar to that produced by structure in the envelope near an outer mass fraction of $10^{-5.5}$.

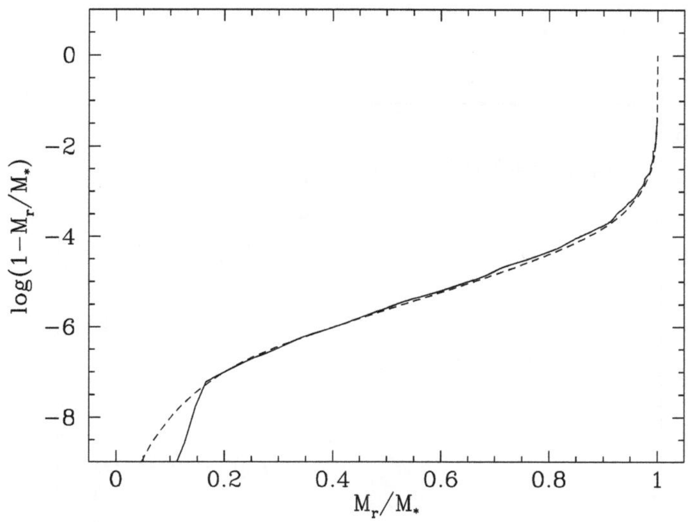

Figure 2. The mapping between points in the core (x-axis) and those in the envelope (y-axis) which produce similar mode trapping for moderate to high overtone modes in a DBV model. The solid line is the result of direct numerical fitting, and the dashed line is the analytical prediction with the inner and outer turning points for each mode fixed at the center and the surface of the star, respectively.

One practical consequence of this core/envelope symmetry is the possibility that we can mis-interpret real structure in the core of a DBV white dwarf by using a model with extra structure at certain locations in the envelope, and vice versa. For example, in Figure 3 we show how the locations of the two chemical transition zones in the model-fit of Fontaine & Brassard (2002) correspond *exactly* to the theoretically expected features at symmetric locations in the core.

We verified this potential problem empirically by cross-fitting the models of Fontaine & Brassard (2002) and Metcalfe (2003)—attempting to match the calculated pulsation periods from one model using the other model to do the fitting. The level of the residuals from these model-to-model comparisons implies that the model-to-observation residuals are dominated by structural uncertainties in the current generation of models, regardless of which model is used to do the fitting.

4. 'BREAKING' THE SYMMETRY

Fortunately, the core/envelope symmetry is only approximate, and there are many ways in which it might be broken. For example, due to the different physical processes that produce them, the generic shape expected for the C/O profile in the core will be different from the expected shape of the He/C profile

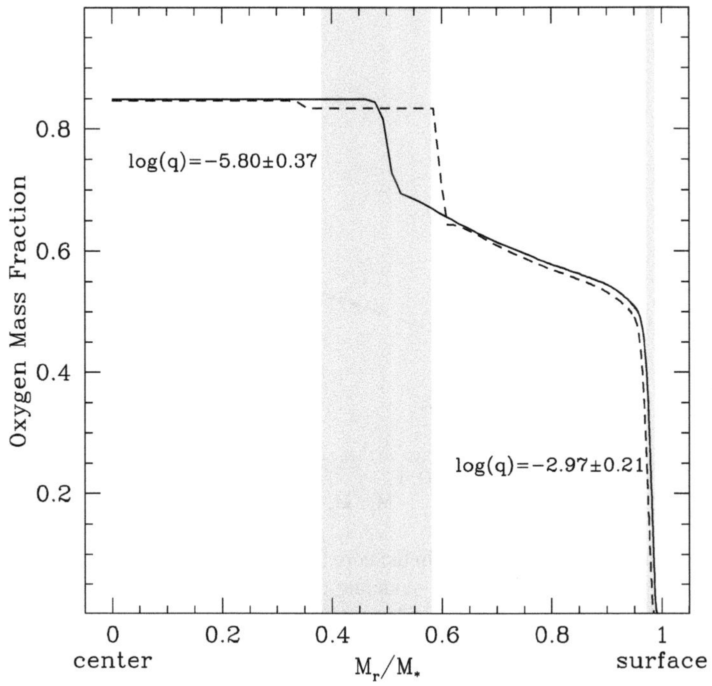

Figure 3. Theoretical internal oxygen profiles for a 0.65 $M_\odot$ white dwarf model produced with standard semiconvective mixing (solid line) and with complete mixing in the overshooting region (dashed line) during central helium burning. The two shaded areas show the regions of the core where perturbations to the Brunt-Väisälä frequency can mimic perturbations in the envelope at values of $\log q$ corresponding to those derived by Fontaine & Brassard (2002) for GD 358.

in the envelope. The reflected perturbations to the Brunt-Väisälä frequency will have different amplitudes and shapes than the original perturbations; in principle we should be able to distinguish between the two (see Figure 4).

5. CONCLUSIONS

Our main conclusions are: (1) white dwarf g-mode pulsations probe *both* the core and the envelope, (2) there is an inherent symmetry that creates an *ambiguity* between the core and envelope structure in DBVs, (3) continued progress will require that we know *detailed* shapes of the internal chemical profiles (i.e. our method is *sensitive* to these shapes), and (4) for other types of pulsating stars (e.g. DAVs) this symmetry may not pose any practical difficulty.

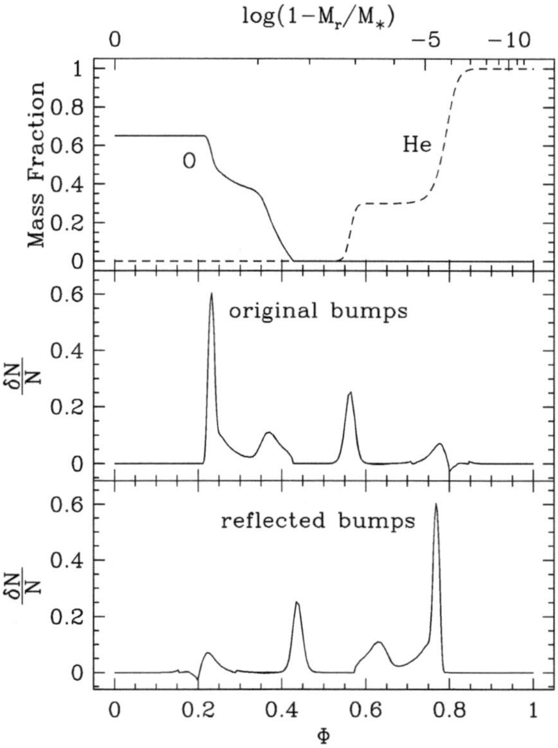

Figure 4. A model that includes theoretical profiles for *both* the C/O and He/C transition zones (top panel), along with the bumps in the Brunt-Väisälä frequency caused by these transition zones (middle panel) and the same bumps reflected through the core/envelope symmetry axis (bottom panel). Note that the deep core transition zone appears to be the most important, followed by the inner He transition.

References

Fontaine, G. & Brassard, P. 2002, *ApJ*, **581**, L33.
Metcalfe T. S., 2003, *ApJ*, **587**, L43.
Montgomery, M. H. et al. 2003, *MNRAS*, **344**, 657.

A SEARCH FOR VARIABILITY IN COOL WHITE DWARF STARS

Terry D. Oswalt, Kyle Johnston, Merissa Rudkin
Florida Institute of Technology

John Bochanski
University of Washington

Justin Schaefer
University of Wyoming

Elizabeth Wennerstrom
Rhodes College

Abstract Preliminary results of a photometric survey of cool white dwarf stars are presented. CCD-based time series observations have been obtained for white dwarfs (WDs) with absolute magnitude $M_v > +15$ and color $0.5 < V - I < 1.3$. Several hours of continuous observations were obtained for each target in an attempt to establish the frequency of microvariability in the temperature regime where collisionally induced absorption due to molecular hydrogen is possible. The minimum time resolution achieved for individual objects is 10-300 sec, depending upon target brightness. In the sample of several dozen cool WDs observed so far, no periodic variability larger than $\sim$2% has been detected. Such objects have good potential to be used as faint ($V > +17$) OIR flux standards for large aperture telescopes.

Keywords: variable stars, white dwarfs

Background

WDs were selected from the Luyten-Giclas (L-G) wide binary catalogs (Luyten 1979; Giclas et al. 1978 and references therein). Figure 1 displays the reduced proper motion diagram ($H_r = m_r + 5log\mu + 5$). The 22 large filled circles are WDs for which we report the results of time series photometry here. Most were selected because they lie near the "knee" in the WD cooling track

E.M. Sion, S. Vennes and H.L. Shipman (eds.), White Dwarfs: Cosmological and Galactic Probes, 85-90.
© 2005 *Springer*.

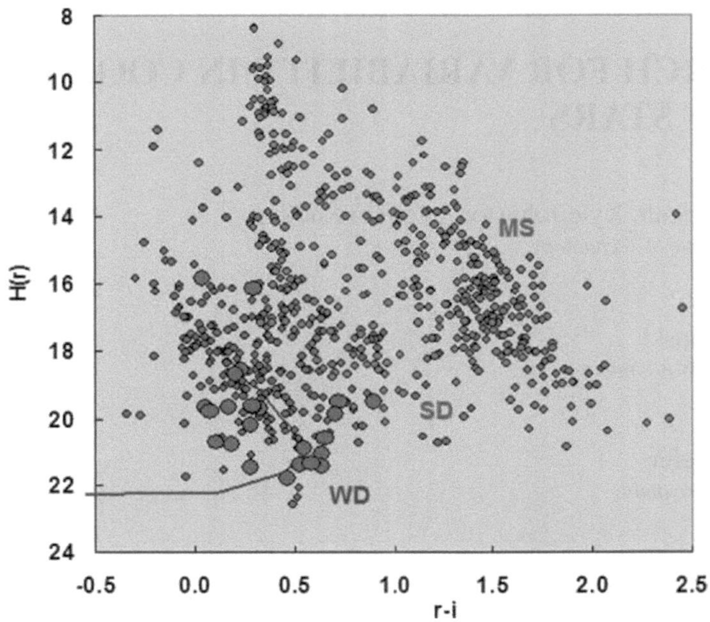

Figure 1. Reduced Proper Motion Diagram for Luyten-Giclas pairs selected for time series survey. The main sequence (MS), subdwarf sequence (SD) and white dwarf cooling track (WD) are indicated. Solid curve indicates Hansen (1999) cooling models for white dwarfs, adjusted for a mean tangential velocity of 73 km/s. White dwarf components observed and reported in this paper are indicated by large filled circles.

(solid curve) computed by Hansen (1999), where the onset of molecular opacity occurs. Several other types of WDs were included for comparison.

1. Observations

Absolute BVRI and unfiltered time-series observations of WD components of wide binaries were obtained with the SARA 0.9-m automated telescope at Kitt Peak in Arizona (see www.saraobservatory.org) by both on-site and remote observers between May 2001 and July 2003. The detector was an Apogee AP7p camera with a back-illuminated SITe SIA 02AB 512K2 CCD. Pixels are 24 microns square, corresponding to ~0.73 arc sec on the sky. Read noise and gain are about 12.2 electrons and 6.1 electrons/ADU, respectively. Integration times were typically 10-300 sec. Sky flats, dark and bias exposures were taken every night. Reductions were performed with standard IRAF packages.

2. Discussion

Table 1 lists the WDs for which we have obtained preliminary unfiltered time series data, along with our absolute BVRI photometry for objects not previously observed by Smith (1997). If scatter in a 2-3 hour normalized light curve exceeded $\sim$2%, a full night of follow-up observations was scheduled. A Discrete Fourier Transform (DFT) was computed for the data set to look for periodic variations smaller than the Nyquist frequency defined by exposure time. Exposure times were chosen to balance the need for relatively high S/N and sensitivity to the range of pulsational periods (100-1000 sec) expected for WDs. Figure 2a and 2b shows representative light curves for LP90-70, a DQ WD, and LP141-14, a DC WD, expressed as normalized ratios of the target WD and the sum of several comparison stars of the same or greater brightness. Among the stars listed in Table 1, no object other than G117-B15A (Figure 2c), which is a well-known DAV (Kepler et al. 2000), exhibits evidence of periodicity with fractional amplitude more than $\sim$2%. Similarly, Nitta et al. (2000) found no pulsations in four cool WDs. Interestingly, LP743-24 and LP813-17 exhibit relatively large random scatter of $>$ 10%. Figure 2d exhibits the serendipitous detection of a partial eclipse involving LP133-373, the distant dM5e companion to LP133-374, a cool DC WD.

3. Null results?

No cool white dwarf star we have observed to date has shown evidence of periodic variability greater than $\sim$2%. However, the time domain has only been sampled at $\sim$10-300 sec time resolution during $\sim$2-6 hour sessions. Moreover, the luminosity and temperature range sampled in the H-R diagram is still quite coarse. Thus, variability at amplitudes, time scales or temperature regimes outside those searched so far cannot be ruled out. Apparently random variations of $\sim$10% occurred in two WD components. The search is being extended to WDs in wide binaries that are $\sim$1-2 magnitudes fainter in M_v, selected from the deeper SuperCOSMOS survey (Pokorny et al. 2003).

Acknowledgments

Support from the NSF to Florida Tech through the SARA Research Experiences for Undergraduates program (AST-0097616) and the Stellar Astronomy and Astrophysics program (AST-0206225) is gratefully acknowledged. This work was also partially supported by NASA AISR subcontract Y701296 through the University of Arizona and the NASA Graduate Student Research Program (NGT5-50450).

References

Giclas, H. L., Burnham, R., & Thomas, N. G. 1978, *Lowell Observatory Bulletin*, **8**, 89.

Hansen, B. M. S. 1999, *ApJ*, **520**, 680.

Kepler, S. O., Mukadam, A., Winget, D. E., Nather, R. E., Metcalfe, T. S., Reed, M. D., Kawaler, S. D., & Bradley, P. A. 2000, *ApJL*, **534**, L185.

Luyten, W.J. 1979, LHS catalogue: A Catalogue of Stars with Proper Motions Exceeding 0"5 Annually, Minneapolis: University of Minnesota, 2nd ed.

Nitta, A., Mukadam, A., Winget, D. E., Kanaan, A., Kleinman, S. J., Kepler, S. O., & Montgomery, M. H. 2000, ASP Conf. Ser. 203: IAU Colloq. 176: The Impact of Large-Scale Surveys on Pulsating Star Research, 525.

Pokorny, R. S., Jones, H. R. A., & Hambly, N. C. 2003, *A&A*, **397**, 575.

Smith, J.A. 1997, Ph.D. thesis, Florida Institute of Technology.

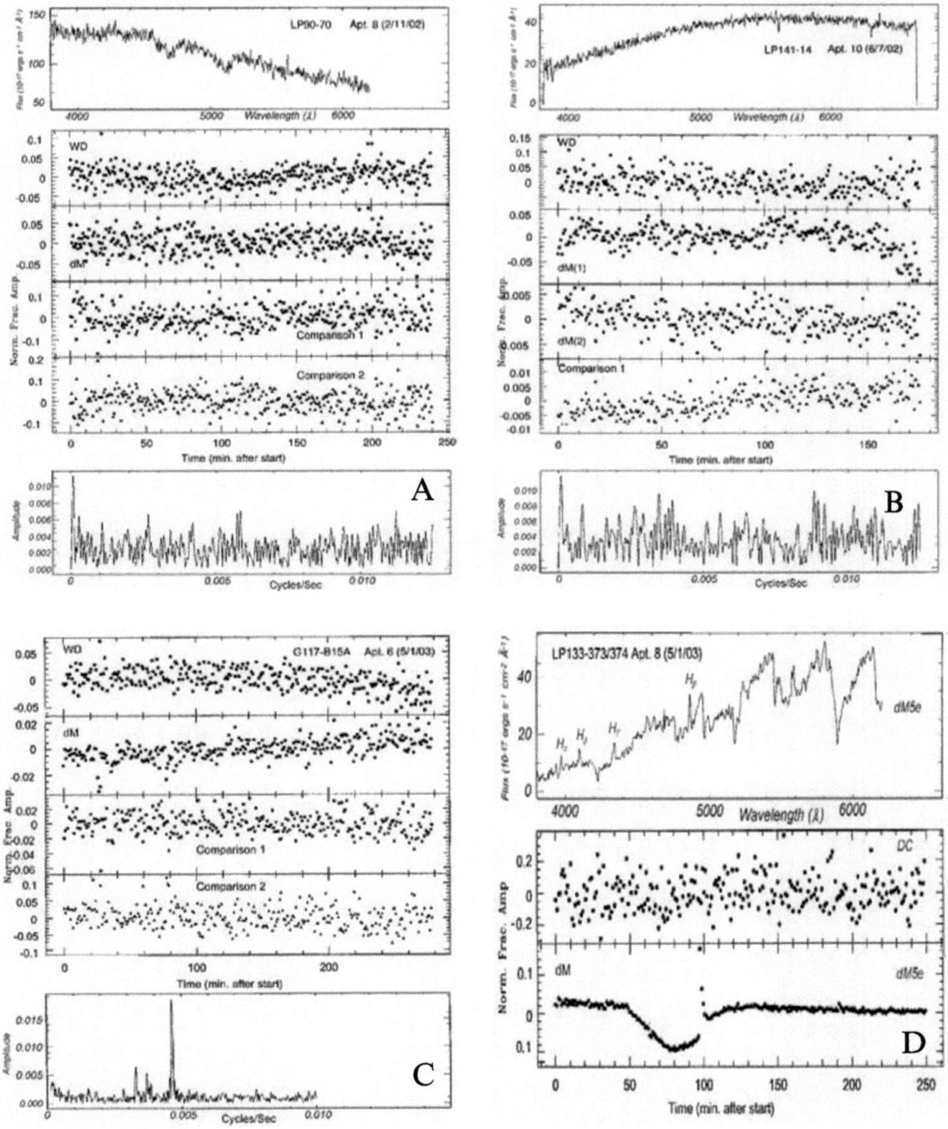

Figure 2. Representative normalized light curves, DFT spectra and optical spectra for (A) LP90-70, a DQ WD; (B) LP141-14, a cool DC WD; (C) G117-B15A a known ZZ Ceti type DAV WD; and (D) the dM5e component LP133-373. See text for details.

Table 1: Time Series Observations of Cool White Dwarf Stars.

Name	R.A. (h m s)	Dec. ° ' "	V	$B-V$	$V-I$	$R-I$	μ ("/yr)	Spec	H_r	Date	Exp.(s) (s)	Δt (s)	N_{obs}	Notes
LP242-12	00 52 30	38 32 00	18.29	0.81	0.59	0.28	0.50	DC	21.47	07/26/01	60; 90	...	...	...
G87-29	07 06 52	37 45 24	15.60	0.38	0.48	0.23	0.44	DQ8	18.57	05/21/02	60	4940	61	...
LP90-70	08 55 40	60 28 18	16.34	0.28	0.24	0.07	0.53	DQ?	19.80	02/11/02	30	14403	362	...
G117-B15A	09 21 13	35 29 48	15.46	0.28	0.22	0.30	0.14	DAV4	16.15	03/21/03	40	16671	295	known DAV; P=215s
LP488-19	09 37 42	09 21 01	18.09	0.92	1.40	0.72	0.26	DC	19.49	02/12/02	30	12523	217	1hr time gap in middle
LP791-55	10 43 30	−18 50 00	15.49	0.56	1.16	0.64	1.98	DQ	21.46	02/13/02	10	...	...	dM companion too close
"	"	"	"	"	"	"	"	"	"	05/19/02	20	...	...	...
LP129-587	11 48 48	54 28 00	16.65	0.39	0.17	0.20	0.25	DA5	18.67	02/13/02	30	3869	93	...
LP96-65	13 08 30	61 34 00	20.20	...	1.15	0.47	0.25	DQ	21.82	05/28/03	300	7489	21	WD2 is C1
LP96-66	13 08 30	61 34 00	19.17	1.13	0.70	0.58	0.29	DC	21.36	05/28/03	300	7489	21	WD1 is target
LP497-30	13 11 30	11 14 00	18.65	1.06	0.67	0.29	0.24	DC	20.17	05/28/03	180	...	...	...
LP617-35	13 17 42	−02 08 00	18.88	1.91	1.23	0.71	0.20	DC	19.87	05/26/03	300	7622	25	very cool WD
LP40-159	13 29 48	72 20 00	18.99	0.58	0.40	0.05	0.16	DA	19.66	05/21/02	180	14099	73	...
LP798-13	13 34 18	−16 04 00	15.33	-0.06	-0.05	0.04	0.12	DA2	15.82	05/02/03	60; 50	10453	137	...
LP380-5	13 46 00	23 52 00	15.46	1.26	0.94	0.55	1.48	DC9	20.92	05/22/01	30	17734	435	...
LP133-374	14 02 24	50 36 00	18.43	0.84	1.35	0.90	0.20	DC:	19.48	05/01/03	45	14935	245	$\sigma > 10\%$
LP176-60	15 33 12	46 59 00	18.23	1.12	1.33	0.64	0.51	DC9	21.07	02/12/02	30	4712	117	very cool WD
LP743-24	15 39 12	−13 47 00	18.74	0.64	0.81	0.29	0.19	DA	19.62	05/27/03	300	4135	14	$\sigma > 10\%$
LP101-16	16 33 30	57 15 00	15.05	0.55	0.49	0.18	1.62	DQ8	20.80	05/24/00	60	...	...	...
"	"	"	"	"	"	"	"	"	"	05/20/02	50	...	...	...
"	"	"	"	"	"	"	"	"	"	03/20/03	70	...	...	...
LP332-27	17 27 36	29 19 00	...	...	...	...	0.25	DA	...	07/24/01	60	6211	83	LP332-28 not in FOV
"	"	"	...	...	...	...	"	"	...	06/06/02	60	...	...	...
"	"	"	...	...	...	...	"	"	...	07/20/02	50	...	...	...
"	"	"	...	...	...	...	"	"	...	05/03/03	80	...	...	...
LP141-14	18 56 30	53 27 00	...	...	..	...	0.23	DC	...	06/07/02	30	10503	261	triple: B=C1, C=C2, both dM
LP813-17	19 43 11	−17 26 00	18.18	0.73	0.73	0.29	0.24	DC	19.64	05/26/01	90	5708	58	...
"	"	"	"	"	"	"	"	"	"	05/29/01	120	...	...	...
VBs11	20 54 06	-05 03 00	16.71	1.16	1.36	0.66	0.82	DC9	20.58	07/25/01	60	...	...	composite?
G24-9	20 11 32	06 32 30	15.78	0.34	0.62	0.33	0.70	DQ7	19.71	05/22/02	60	9288	110	composite, exclipsing B=C3
"	"	"	"	"	"	"	"	"	"	07/08/03	90	...	...	...
LP698-4	21 30 48	−06 24 00	18.52	0.43	0.50	0.11	0.33	DC	20.72	07/25/01	60	...	...	...
LP701-69	23 01 42	−07 17 24	17.20	0.27	0.38	0.17	0.34	DB	19.64	07/25/01	60	5606	76	...
LP701-70	23 01 42	−07 17 24	19.39	1.49	1.17	0.53	0.34	DC	21.41	07/25/01	60	...	...	LP701-70 not in FOV
LHS4033	23 49 55	−03 10 08	...	...	...	...	0.66	DA	...	07/24/01	60	11356	162	...
"	"	"	...	...	...	...	"	"	...	09/26/03	90	...	...	...
LP77-56	23 51 36	65 05 42	...	...	...	...	0.41	DC	...	10/17/01	120	19460	149	WD2; C1 value is for WD2
"	"	"	...	...	...	...	"	"	...	10/20/01	120	...	...	...
LP77-57	23 51 36	65 05 42	...	...	...	...	0.41	DQ	...	10/17/01	120	19460	149	WD1; target value is for WD1
"	"	"	...	...	...	...	"	"	...	10/20/01	120	...	...	...

STUDY OF THE GALACTIC BULGE USING NEW PLANETARY NEBULAE DISCOVERED ON THE AAO/UKST Hα SURVEY

A.E.J. Peyaud
Macquarie University
Strasbourg Observatory
alanp@ics.mq.edu.au

Q.A. Parker
Macquarie University
Anglo-Australian Observatory
qap@ics.mq.edu.au

A. Acker
Strasbourg Observatory
acker@newb6.u-strasbg.fr

Abstract The AAO/UKST Hα survey is revealing a substantial new population of Planetary Nebulae (PNe) over the southern Galactic Plane including $\sim$ 550 in the Bulge. Such objects are excellent tracers for Bulge dynamical studies, are less affected by metallicity bias and have strong emission lines facilitating accurate velocities. We present preliminary results on plans to exploit these significant new Bulge PNe. Until now the relatively small numbers previously known has limited their potential as tracers of galactic structure and as windows to stellar evolution in the critical phase between PNe and White Dwarf (e.g. Zijlstra et al. 1997; Durand et al.1996, 1998; Beaulieu et al. 1999).

Keywords: Galactic bulge, planetary nebulae

1. New PNe from the AAO/UKST Hα Survey & their Spectroscopic follow-up

The AAO UK Schmidt Telescope has undertaken a high resolution, narrow-band, Hα survey of the Southern Galactic Plane (Parker et al 2003a). This survey has revealed a substantial new population of PNe but with a particular rich-

E.M. Sion, S. Vennes and H.L. Shipman (eds.), White Dwarfs: Cosmological and Galactic Probes, 91-93.
© 2005 *Springer.*

ness in the Bulge (Parker et al. 2003b). The MASH (Macquarie/AAO/Strasbourg Hα) PNe project, set-up to exploit this resource, is significantly impacting PNe research and will more than double Galactic PNe. Candidates are found using difference imaging (Bond et al. 2002) applied to the matching Hα/SR images. SExtractor (Bertin et al. 1995) is then used to objectively extract possible PNe from the difference maps. Spectroscopic follow-up on 6dF (UKST) provides confirmation. Some new Bulge PNe are extended (and presumably evolved) but a large fraction are more compact than previous visually discovered samples and may relate to less evolved PNe.

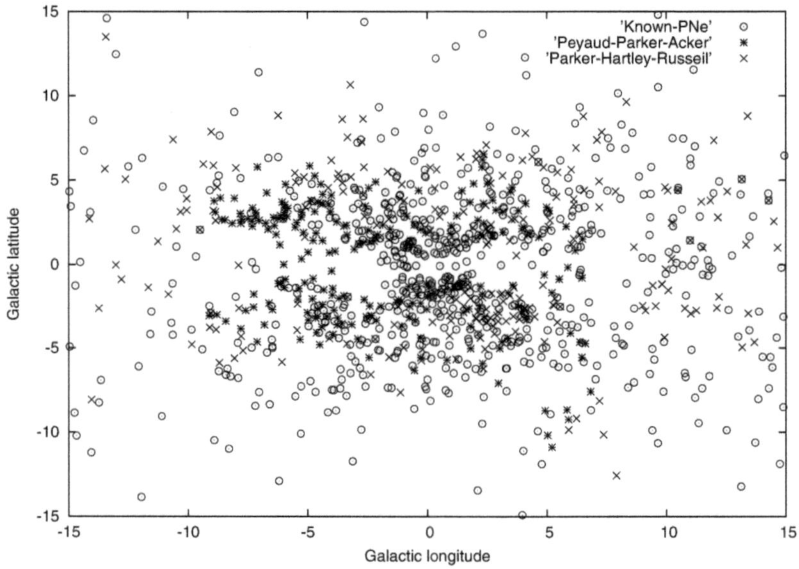

Figure 1. Bulge PNe distribution (includes known PNe & new confirmed MASH objects)

2. The Bulge PNe population & Probes of Bulge Dynamics

MASH has already discovered $\sim$ 550 new, generally compact, Bulge PNe bringing the total to $\sim$1200. This new sample should provide further insight into the stellar population of the Bulge (e.g. Acker et al. 1996). N-body models of bar forming instabilities of galactic disks provide theoretical predictions of the resulting bar/bulges. These can be directly tested against observed dynamical data from our new overall Bulge PNe sample. In particular the limited number of Bulge PNe known previously $\sim$650 has restricted their potential to test Bulge dynamical models. Once our Bulge coverage is complete this key, new, homogenously derived sample will provide sufficient particles for detailed comparisons with the latest dynamical models. The complexity of the

Milky Way Bulge in terms of its dynamics and stellar populations is daunting, yet it is a promising to study as it contains key information on fundamental questions regarding formation and evolution of the Galaxy. Within the framework of the current standard picture of galaxy formation, a specific question we wish investigate is the proposition that the Galactic bulge was the largest of the primordial fragments that now form the Milky Way, giving a centre of nucleation around which the remainder of the Milky Way could grow. We also wish to investigate the importance of bulges in determining evolutionary history of spiral galaxies. The Bulge offers the unique opportunity to obtain an accurate phase-space portrait and metallicity distribution of a Bulge population which can be used to address these questions (Ibata & Gilmore 1995). A test for this accretion scenario would be to find a homogeneous velocity distribution as a function of chemistry.

References

Acker A. et al., 1996, Strasbourg-ESO Catalogue of Galactic PN (Garching: ESO).

Beaulieu, S., et al. 2000, *ApJ*, **515**, 610.

Bertin, E. & Arnouts, S., 1995, *A&A Suppl. Ser.*, **117**, 393.

Bond I., et al. 2001, *MNRAS*, **327**, 868.

Durand, S., Dejonghe, H., Acker, A. 1996, *A&A*, **310**, 97.

Durand, S., Acker A., Zijlstra, A.A. 1998, *A&A Suppl. Ser.*, **132**, 12.

Ibata, R., & Gilmore, G. 1995, *MNRAS*, **275**, 605.

Parker, Q.A., et al. 2003a, in *ASP Conf.Ser.*, **232**, 38.

Parker, Q.A., et al. 2003b IAU Symp. 209, in "Planetary Nebulae: Their Evolution & Role in the Universe", eds. Sun Kwok, Michael Dopita & Ralph Sutherland, 25.

Zijlstra, A.A., Acker, A., & Walsh, J.R. 1997, *A&A Suppl. Ser.*, **125**, 289.

SECTION II. TYPE IA SUPERNOVAE

TYPE IA SUPERNOVAE AND COSMOLOGY

Review paper

Alexei V. Filippenko
Department of Astronomy, University of California,
Berkeley, CA 94720-3411, USA
alex@astro.berkeley.edu

Abstract I discuss the use of Type Ia supernovae (SNe Ia) for cosmological distance determinations. Low-redshift SNe Ia ($z \lesssim 0.1$) demonstrate that the Hubble expansion is linear with $H_0 = 72 \pm 8$ km s^{-1} Mpc^{-1}, and that the properties of dust in other galaxies are generally similar to those of dust in the Milky Way. The light curves of high-redshift ($z = 0.3$–1) SNe Ia are stretched in a manner consistent with the expansion of space; similarly, their spectra exhibit slower temporal evolution (by a factor of $1 + z$) than those of nearby SNe Ia. The measured luminosity distances of SNe Ia as a function of redshift have shown that the expansion of the Universe is currently accelerating, probably due to the presence of repulsive dark energy such as Einstein's cosmological constant (Λ). From about 200 SNe Ia, we find that $H_0 t_0 = 0.96 \pm 0.04$, and $\Omega_\Lambda - 1.4\Omega_M = 0.35 \pm 0.14$. Combining our data with the results of large-scale structure surveys, we find a best fit for Ω_M and Ω_Λ of 0.28 and 0.72, respectively — essentially identical to the recent *WMAP* results (and having comparable precision). The sum of the densities, ~ 1.0, agrees with extensive measurements of the cosmic microwave background radiation, including *WMAP*, and coincides with the value predicted by most inflationary models for the early Universe: the Universe is flat on large scales. A number of possible systematic effects (dust, supernova evolution) thus far do not seem to eliminate the need for $\Omega_\Lambda > 0$. However, during the past few years some very peculiar low-redshift SNe Ia have been discovered, and we must be mindful of possible systematic effects if such objects are more abundant at high redshifts. Recently, analyses of SNe Ia at $z = 1.0$–1.7 provide further support for current acceleration, and give tentative evidence for an early epoch of deceleration. The dynamical age of the Universe is estimated to be 13.1 ± 1.5 Gyr, consistent with the ages of globular star clusters and with the *WMAP* result of 13.7 ± 0.2 Gyr. According to the most recent data sets, the SN Ia rate at $z > 1$ is several times greater than that at low redshifts, presumably because of higher star formation rates long ago. Moreover, the typical delay time from progenitor star formation to SN Ia explosion appears to be substantial, ~ 3 Gyr. Current projects include the search for additional SNe Ia at $z > 1$ to confirm the early deceleration, and the measurement of a few hundred SNe Ia at $z = 0.2$–0.8 to more accurately determine the equation-of-state parameter of the dark energy,

E.M. Sion, S. Vennes and H.L. Shipman (eds.), White Dwarfs: Cosmological and Galactic Probes, 97-133.
© 2005 *Springer.*

$w = P/(\rho c^2)$, whose value is now constrained by SNe Ia to be in the range $-1.48 \lesssim w \lesssim -0.72$ at 95% confidence.

Keywords: cosmological parameters, dark energy, distance scale, supernovae

1. Introduction

Supernovae (SNe) come in two main observational varieties (see Filippenko 1997b for a review). Those whose optical spectra exhibit hydrogen are classified as Type II, while hydrogen-deficient SNe are designated Type I. SNe I are further subdivided according to the detailed appearance of the early-time spectrum: SNe Ia are characterized by strong absorption near 6150 Å now attributed to Si II, SNe Ib lack this feature but instead show prominent He I lines, and SNe Ic have neither the Si II nor the He I lines (at least not strong ones). SNe Ia are believed to result from the thermonuclear disruption of carbon-oxygen white dwarfs, while SNe II come from core collapse in massive supergiant stars. The latter mechanism probably produces most SNe Ib/Ic as well, but the progenitor stars previously lost their outer layers of hydrogen or even helium, through either winds or mass transfer onto a companion star.

It has long been recognized that SNe Ia may be very useful distance indicators for a number of reasons; see Branch & Tammann (1992), Branch (1998), and references therein. (1) They are exceedingly luminous, with peak M_B averaging -19.0 mag if $H_0 = 72$ km s^{-1} Mpc^{-1}. (2) "Normal" SNe Ia have small dispersion among their peak absolute magnitudes ($\sigma \lesssim 0.3$ mag). (3) Our understanding of the progenitors and explosion mechanism of SNe Ia is on a reasonably firm physical basis. The results of recent models give good fits to the observed spectra and light curves, providing confidence that we are not far off the mark. (4) Little cosmic evolution is expected in the peak luminosities of SNe Ia, and it can be modeled. (5) One can perform *local* tests of various possible complications and evolutionary effects by comparing nearby SNe Ia in different environments (elliptical galaxies, bulges and disks of spirals, galaxies having different metallicities, etc.).

Research on SNe Ia in the 1990s has demonstrated their enormous potential as distance indicators. Although there are subtle effects that must indeed be taken into account, it appears that SNe Ia provide among the most accurate values of H_0, q_0 (the deceleration parameter), Ω_M (the matter density), and Ω_Λ [the cosmological constant, $\Lambda c^2/(3H_0^2)$].

For more than a decade there have been two major teams involved in the systematic investigation of high-redshift SNe Ia for cosmological purposes. The "Supernova Cosmology Project" (SCP) is led by Saul Perlmutter of the Lawrence Berkeley Laboratory, while the "High-Z supernova search Team" (HZT) is led by Brian Schmidt of the Mt. Stromlo and Siding Springs Observatories. I have been privileged to work with both of these teams (see Filippenko

2001 for a personal account), but my primary allegiance is now with the HZT. A few years ago, the HZT split into two overlapping subsets: the "Higher-Z Supernova Search Team" led by Adam Riess of the Space Telescope Science Institute, and the ESSENCE team ("Equation of State: SupErNovae trace Cosmic Expansion") led by Christopher Stubbs of Harvard University. Other groups have recently formed to conduct similar studies, such as the supernova team of the Canada-France-Hawaii Telescope Legacy Survey. An outgrowth of the SCP, the very large SNAP (SuperNova/Acceleration Probe) collaboration is planning a future space satellite that will be largely devoted to using SNe Ia for cosmology.

2. Homogeneity and Heterogeneity

Until the mid-1990s, the traditional way in which SNe Ia were used for cosmological distance determinations was to assume that they are perfect "standard candles" and to compare their observed peak brightness with that of SNe Ia in galaxies whose distances had been independently determined (e.g., with Cepheid variables). The rationale was that SNe Ia exhibit relatively little scatter in their peak blue luminosity ($\sigma_B \approx$ 0.4–0.5 mag; Branch & Miller 1993), and even less if "peculiar" or highly reddened objects were eliminated from consideration by using a color cut. Moreover, the optical spectra of SNe Ia are usually rather homogeneous, if care is taken to compare objects at similar times relative to maximum brightness (Riess et al. 1997, and references therein). Over 80% of all SNe Ia discovered through the early 1990s were "normal" (Branch, Fisher, & Nugent 1993).

From a Hubble diagram constructed with unreddened, moderately distant SNe Ia ($z \lesssim 0.1$) for which peculiar motions are small and relative distances (given by ratios of redshifts) are accurate, Vaughan et al. (1995) find that

$$\langle M_B(\mathrm{max})\rangle = (-19.74 \pm 0.06) + 5\log{(H_0/50)}\ \mathrm{mag}. \quad (1)$$

In a series of papers, Sandage et al. (1996) and Saha et al. (1997) combine similar relations with *Hubble Space Telescope (HST)* Cepheid distances to the host galaxies of seven SNe Ia to derive $H_0 = 57 \pm 4$ km s^{-1} Mpc^{-1}.

Over the past two decades it has become clear, however, that SNe Ia do *not* constitute a perfectly homogeneous subclass (e.g., Filippenko 1997a,b). In retrospect this should have been obvious: the Hubble diagram for SNe Ia exhibits scatter larger than the photometric errors, the dispersion actually *rises* when reddening corrections are applied (under the assumption that all SNe Ia have uniform, very blue intrinsic colors at maximum; van den Bergh & Pazder 1992; Sandage & Tammann 1993), and there are some significant outliers whose anomalous magnitudes cannot be explained by extinction alone.

Spectroscopic and photometric peculiarities have been noted with increasing frequency in well-observed SNe Ia. A striking case is SN 1991T; its pre-maximum spectrum did not exhibit Si II or Ca II absorption lines, yet two months past maximum brightness the spectrum was nearly indistinguishable from that of a classical SN Ia (Filippenko et al. 1992b; Ruiz-Lapuente et al. 1992; Phillips et al. 1992). The light curves of SN 1991T were slightly broader than the SN Ia template curves, and the object was probably somewhat more luminous than average at maximum. Another well-observed, peculiar SNe Ia is SN 1991bg (Filippenko et al. 1992a; Leibundgut et al. 1993; Turatto et al. 1996). At maximum brightness it was subluminous by 1.6 mag in V and 2.5 mag in B, its colors were intrinsically red, and its spectrum was peculiar (with a deep absorption trough due to Ti II). Moreover, the decline from maximum was very steep, the I-band light curve did not exhibit a secondary maximum like normal SNe Ia, and the velocity of the ejecta was unusually low. The photometric heterogeneity among SNe Ia is well demonstrated by Suntzeff (1996) with objects having excellent $BVRI$ light curves.

3. Cosmological Uses: Low Redshifts

Although SNe Ia can no longer be considered perfect "standard candles," they are still exceptionally useful for cosmological distance determinations. Excluding those of low luminosity (which are hard to find, especially at large distances), most of the nearby SNe Ia that had been discovered through the early 1990s were *nearly* standard (Branch et al. 1993; but see Li et al. 2001b for more recent evidence of a higher intrinsic peculiarity rate). Also, after many tenuous suggestions (e.g., Pskovskii 1977, 1984; Branch 1981), Phillips (1993) found convincing evidence for a correlation between light-curve shape and the luminosity at maximum brightness by quantifying the photometric differences among a set of nine well-observed SNe Ia, using a parameter [$\Delta m_{15}(B)$] that measures the total drop (in B magnitudes) from B-band maximum to $t = 15$ days later. In all cases the host galaxies of his SNe Ia have accurate relative distances from surface brightness fluctuations or from the Tully-Fisher relation. The intrinsically bright SNe Ia clearly decline more slowly than dim ones, but the correlation is stronger in B than in V or I.

Using SNe Ia discovered during the Calán/Tololo survey ($z \lesssim 0.1$), Hamuy et al. (1995, 1996b) refine the Phillips (1993) correlation between peak luminosity and $\Delta m_{15}(B)$. Apparently the slope is steep only at low luminosities; thus, objects such as SN 1991bg skew the slope of the best-fitting single straight line. Hamuy et al. reduce the scatter in the Hubble diagram of normal, unreddened SNe Ia to only 0.17 mag in B and 0.14 mag in V; see also Tripp (1997). Yet another parameterization is the "stretch" method of Perlmutter et al. (1997) and Goldhaber et al. (2001): the B-band light curves of SNe Ia appear nearly identical when expanded or contracted temporally by a factor s, where the value of s varies among objects. In a similar but distinct effort, Riess, Press, & Kirshner (1995) show that the luminosity of SNe Ia correlates with the detailed *shape* of the overall light curve.

By using light-curve shapes measured through several different filters, Riess, Press, & Kirshner (1996a) extend their analysis and objectively eliminate the effects of interstellar extinction (as is also now done with the Δm_{15} method; Phillips et al. 1999). A SN Ia that has an unusually red $B - V$ color at maximum brightness is assumed to be *intrinsically* subluminous if its light curves rise and decline quickly, or of normal luminosity but significantly *reddened* if its light curves rise and decline more slowly. With a set of 20 SNe Ia from the Calán/Tololo sample (Hamuy et al. 1996c) and the Harvard/Smithsonian Center for Astrophysics (CfA) follow-up program (Riess et al. 1999a), Riess et al. (1996a) show that the dispersion decreases from 0.52 mag to 0.12 mag after application of this "multi-color light-curve shape" (MLCS) method. The results from an expanded set of nearly 50 SNe Ia indicate that the dispersion decreases from 0.44 mag to 0.15 mag (Figure 1). The resulting Hubble constant is 65 ± 2 (statistical) ± 7 (systematic) km s^{-1} Mpc^{-1}, with an additional systematic and zero-point uncertainty of ± 5 km s^{-1} Mpc^{-1}. (Re-calibrations of the Cepheid distance scale, and other recent refinements, lead to a best estimate of $H_0 = 72 \pm 8$ km s^{-1} Mpc^{-1}, where the error bar includes both statistical and systematic uncertainties; Freedman et al. 2001.) Saha et al. (2001) still argue for a longer distance scale, with $H_0 \approx 60$ km s^{-1} Mpc^{-1}, using different choices in measuring the supernova distances (Parodi et al. 2000) and the Cepheid distances; see the discussions by Gibson et al. (2000) and Jha (2002).

Apart from the controversy regarding their true peak luminosity, low-redshift SNe Ia provide the best evidence that the Hubble flow is linear (Riess et al. 1996a), with deviations from linearity well-explained by random peculiar velocities and the local flow field (Riess et al. 1997; Jha 2002). Phillips et al. (1999) and Riess et al. (1996a) further argue that the dust affecting SNe Ia is *not* of circumstellar origin, and show quantitatively that the extinction curve in external galaxies typically does not differ from that in the Milky Way (cf. Branch & Tammann 1992, but see Tripp 1998).

The advantage of systematically correcting the luminosities of SNe Ia at high redshifts rather than trying to isolate "normal" ones seems clear in view of evidence that the luminosity of SNe Ia may be a function of stellar population. If the most luminous SNe Ia occur in young stellar populations (e.g., Hamuy et al. 1996a, 2000; Branch, Romanishin, & Baron 1996; Ivanov, Hamuy, & Pinto 2000), then we might expect the mean peak luminosity of high-z SNe Ia to differ from that of a local sample. Alternatively, the use of Cepheids (Population I objects) to calibrate local SNe Ia can lead to a zero-point that is too luminous. On the other hand, as long as the physics of SNe Ia is essentially

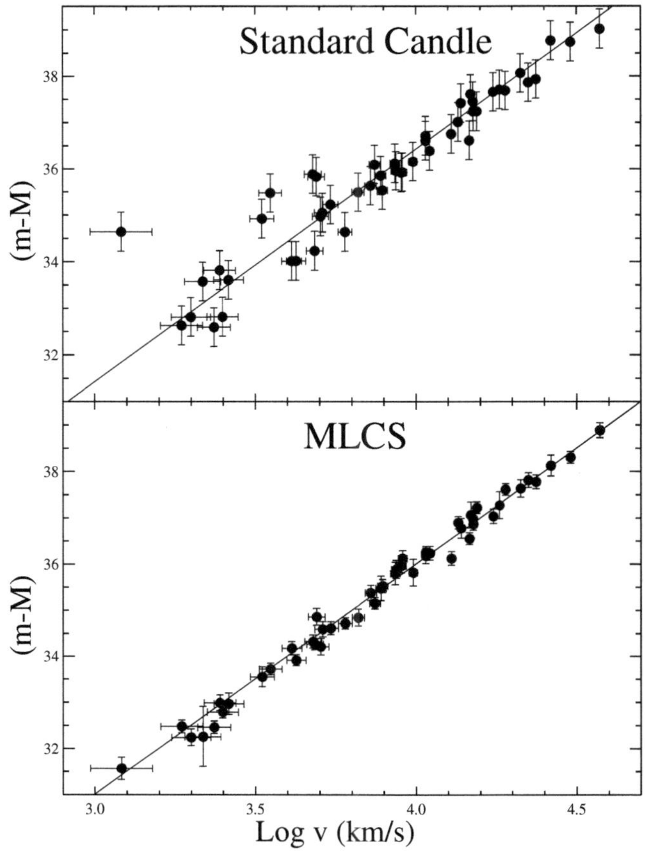

Figure 1. Hubble diagrams for SNe Ia (A. G. Riess 2001, private communication) with velocities (km s^{-1}) in the $COBE$ rest frame on the Cepheid distance scale. The ordinate shows distance modulus, $m - M$ (mag). Top: The objects are assumed to be $standard\ candles$ and there is no correction for extinction; the result is $\sigma = 0.42$ mag and $H_0 = 58 \pm 8$ km s^{-1} Mpc^{-1}. Bottom: The same objects, after correction for reddening and intrinsic differences in luminosity. The result is $\sigma = 0.15$ mag and $H_0 = 65 \pm 2$ (statistical) km s^{-1} Mpc^{-1}, subject to changes in the zero-point of the Cepheid distance scale. (Indeed, the latest results with SNe Ia favor $H_0 = 72$ km s^{-1} Mpc^{-1}.)

the same in young stellar populations locally and at high redshift, we should be able to adopt the luminosity correction methods (photometric and spectroscopic) found from detailed studies of low-z SNe Ia.

In the past few years, many nearby SNe have been found by industrious amateur astronomers including R. Arbour, M. Armstrong, T. Boles, T. Puckett, M. Schwartz, and others. The "Nearby Supernova Factory" run by G. Aldering's team at the Lawrence Berkeley National Laboratory is also responsible for many discoveries, when it is conducting a search. However, about half of all reported nearby SNe in the last half-decade were discovered by my team's Lick Observatory Supernova Search (LOSS) conducted with the 0.76-m Katzman Automatic Imaging Telescope (KAIT; Li et al. 2000; Filippenko et al. 2001; Filippenko 2003; see http://astro.berkeley.edu/~bait/kait.html). During LOSS, CCD images are taken of $\gtrsim 1000$ galaxies per night and compared with KAIT "template images" obtained earlier; the templates are automatically subtracted from the new images and analyzed with computer software, and the SN candidates are flagged. The next day, undergraduate students at UC Berkeley examine all candidates, including weak ones, to eliminate star-like cosmic rays, asteroids, and other sources of false alarms. They also glance at some of the subtracted images to locate SNe that might be near bright, poorly subtracted stars or galactic nuclei. LOSS discovered 20 SNe in 1998, 40 in 1999, 38 in 2000, 68 in 2001, 82 in 2002, and 95 in 2003, making it by far the world's leading search for nearby SNe.

Spectroscopic classifications of these and other low-redshift SNe are provided by a number of groups (see the *IAU Circulars*), including our own observations with the Lick 3-m and Keck telescopes, the CfA SN monitoring campaign, the Texas/McDonald Observatory group, the Asiago team, the European Research and Training Network, the Australian National University team, etc. The most important objects were photometrically monitored with KAIT through $BVRI$ (and sometimes U) filters (e.g., Li et al. 2001a, 2003b; Modjaz et al. 2001; Leonard et al. 2002a,b; Foley et al. 2003), and unfiltered follow-up observations (e.g., Matheson et al. 2001) were made of almost all of them during the course of LOSS. Note that recently, KAIT has also been used to automatically monitor the optical afterglows of gamma-ray bursts (Li et al. 2003a,c; Matheson et al. 2003).

This growing sample of well-observed SNe Ia allows us to more precisely calibrate the distance determinations, as well as to look for correlations between the observed properties of the SNe and their environment (Hubble type of host galaxy, metallicity, stellar population, etc.). Jha (2002) present $UBVRI$ follow-up observations from the CfA SN monitoring campaign of 44 nearby SNe Ia (including many discovered by KAIT), from which they derive MLCS2k2, an updated light-curve fitting technique that includes U-band templates (critical for SN Ia observations at high redshift). Recently, progress has also been made in the near-infrared, with Krisciunas et al. (2001, 2003, 2004) presenting JHK light curves of nearby SNe Ia. These have enabled Krisciunas, Phillips, & Suntzeff (2004) to construct near-infrared Hubble diagrams of SNe Ia, with significantly less effect from dust extinction along the line of sight. Their results suggest that SNe Ia are much closer to *standard* candles in the near-infrared than in the optical, with little dependence of light-curve shape on near-infrared luminosity.

4. Cosmological Uses: High Redshifts

These same techniques can be applied to construct a Hubble diagram with high-redshift SNe Ia, from which the value of $q_0 = (\Omega_M/2) - \Omega_\Lambda$ can be determined. With enough objects spanning a range of redshifts, we can measure Ω_M and Ω_Λ independently (e.g., Goobar & Perlmutter 1995). Contours of peak apparent R-band magnitude for SNe Ia at two redshifts have different slopes in the Ω_M–Ω_Λ plane, and the regions of intersection provide the answers we seek.

The Search

Based on the pioneering work of Norgaard-Nielsen et al. (1989), whose goal was to find SNe in moderate-redshift clusters of galaxies, the SCP (Perlmutter et al. 1995a, 1997) and the HZT (Schmidt et al. 1998) devised a strategy that almost guarantees the discovery of many faint, distant SNe Ia "on demand," during a predetermined set of nights. This "batch" approach to studying distant SNe allows follow-up spectroscopy and photometry to be scheduled in advance, resulting in a systematic study not possible with random discoveries. Most of the searched fields are equatorial, permitting follow-up from both hemispheres. The SCP was the first group to convincingly demonstrate the ability to find SNe in batches.

Our approach is simple in principle; see Schmidt et al. (1998) for details, and for a description of our first high-redshift SN Ia (SN 1995K). Pairs of first-epoch images during the nights around new moon are obtained with wide-field cameras on large telescopes (e.g., the Big Throughput Camera on the CTIO 4-m Blanco telescope, and more recently the CTIO Mosaic II, the CFHT 8K and 12K mosaics, and Suprime Cam on Subaru), followed by second-epoch images 3–4 weeks later. (Pairs of images permit removal of cosmic rays, asteroids, and Kuiper-belt objects.) These are compared immediately using well-tested software, and new SN candidates are identified in the second-epoch images (Figure 2). Spectra are obtained as soon as possible after discovery to verify that the objects are SNe Ia and to determine their redshifts. Each team has already found over 200 SNe in concentrated batches, as reported in numerous *IAU Circulars* (e.g., Perlmutter et al. 1995b, 11 SNe with $0.16 \lesssim z \lesssim 0.65$; Suntzeff et al. 1996, 17 SNe with $0.09 \lesssim z \lesssim 0.84$).

Intensive photometry of the SNe Ia commences within a few days after procurement of the second-epoch images; it is continued throughout the ensuing and subsequent dark runs. In a few cases *HST* images are obtained. As expected, most of the discoveries are *on the rise or near maximum brightness.* When possible, the SNe are observed in filters that closely match the redshifted B and V bands; this way, the K-corrections become only a second-order effect (Kim, Goobar, & Perlmutter 1996; Nugent, Kim, & Perlmutter 2002).

Figure 2. Discovery image of SN 1997cj (28 April 1997), along with the template image and an HST image obtained subsequently. The net (subtracted) image is also shown.

Observations through two filters allow us to apply reddening and luminosity corrections (Riess et al. 1996a; Hamuy et al. 1996a,b; Phillips et al. 1999).

Although SNe in the magnitude range 22–22.5 can sometimes be spectroscopically confirmed with 4-m class telescopes, the signal-to-noise ratios are low, even after several hours of integration. Certainly Keck, Gemini, the VLT, Subaru, or Magellan are required for the fainter objects (22.5–24.5 mag). With the largest telescopes, not only can we rapidly confirm a substantial number of candidate SNe, but we can search for peculiarities in the spectra that might indicate evolution of SNe Ia with redshift. Moreover, high-quality spectra allow us to measure the age of a SN: we have developed a method for automatically comparing the spectrum of a SN Ia with a library of spectra from many different epochs in the development of SNe Ia (Riess et al. 1997). Our technique also has great practical utility at the telescope: we can determine the age of a SN "on the fly," within half an hour after obtaining its spectrum. This allows us to decide rapidly which SNe are best for subsequent photometric follow-up, and we immediately alert our collaborators elsewhere.

Results

First, we note that the light curves of high-redshift SNe Ia are broader than those of nearby SNe Ia; the initial indications (Leibundgut et al. 1996; Goldhaber et al. 1997), based on small numbers of SNe Ia, are amply confirmed with the larger samples (Goldhaber et al. 2001). Quantitatively, the amount by which the light curves are "stretched" is consistent with a factor of $1+z$, as expected if redshifts are produced by the expansion of space rather than by "tired light" and other non-expansion hypotheses for the redshifts of objects at cosmological distances. [For non-standard cosmological interpretations of the SN Ia data, see Narlikar & Arp (1997) and Hoyle, Burbidge, & Narlikar (2000).] We also demonstrate this *spectroscopically* at the 2σ confidence level for a single object: the spectrum of SN 1996bj ($z = 0.57$) evolved more slowly than those of nearby SNe Ia, by a factor consistent with $1+z$ (Riess et al. 1997). Observations of SN 1997ex ($z = 0.36$) at three epochs conclusively verify the effects of time dilation: temporal changes in the spectra are slower than those of nearby SNe Ia by roughly the expected factor of 1.36. Although one might be able to argue that something other than universal expansion could be the cause of the apparent stretching of SN Ia light curves at high redshifts, it is much more difficult to attribute apparently slower evolution of spectral details to an unknown effect.

The formal value of Ω_M derived from SNe Ia has changed with time. The SCP published the initial result (Perlmutter et al. 1995a), based on a single object, SN 1992bi at $z = 0.458$: $\Omega_M = 0.2 \pm 0.6 \pm 1.1$ (assuming that $\Omega_\Lambda = 0$). The SCP's analysis of their first seven objects (Perlmutter et al. 1997) suggested a much larger value of $\Omega_M = 0.88 \pm 0.6$ (if $\Omega_\Lambda = 0$) or $\Omega_M = 0.94 \pm 0.3$ (if $\Omega_{\rm total} = 1$). Such a high-density universe seemed at odds with other, independent measurements of Ω_M at that time. However, with the subsequent inclusion of just one more object, SN 1997ap at $z = 0.83$ (the highest then known for a SN Ia; Perlmutter et al. 1998), their estimates were revised back down to $\Omega_M = 0.2 \pm 0.4$ if $\Omega_\Lambda = 0$, and $\Omega_M = 0.6 \pm 0.2$ if $\Omega_{\rm total} = 1$; the apparent brightness of SN 1997ap had been precisely measured with *HST*, so it substantially affected the best fits.

Meanwhile, the HZT published (Garnavich et al. 1998a) an analysis of four objects (three of them observed with *HST*), including SN 1997ck at $z = 0.97$, at that time a redshift record, although they cannot be absolutely certain that the object was a SN Ia because the spectrum is too poor. From these data, the HZT derived that $\Omega_M = -0.1 \pm 0.5$ (assuming $\Omega_\Lambda = 0$) and $\Omega_M = 0.35 \pm 0.3$ (assuming $\Omega_{\rm total} = 1$), inconsistent with the high Ω_M initially found by Perlmutter et al. (1997) but consistent with the revised estimate in Perlmutter et al. (1998). An independent analysis of 10 SNe Ia using the "snapshot" distance method (with which conclusions are drawn from sparsely observed SNe Ia) gave quantitatively similar conclusions (Riess et al. 1998a). However, none of these early data sets carried the statistical discriminating power to detect cosmic acceleration.

The SCP's next results were announced at the 1998 January AAS meeting in Washington, DC. A press conference was scheduled, with the stated purpose of presenting and discussing the then-current evidence for a low-Ω_M universe as published by Perlmutter et al. (1998; SCP) and Garnavich et al. (1998a; HZT). When showing the SCP's Hubble diagram for SNe Ia, however, Perlmutter also pointed out tentative evidence for *acceleration*! He said that the conclusion was uncertain, and that the data were equally consistent with no acceleration; the systematic errors had not yet been adequately assessed. Essentially the same conclusion was given by the SCP in February 1998 during their talks at a major conference on dark matter held near Los Angeles (Goldhaber & Perlmutter 1998).

Although it chose not to reveal them at the same 1998 January AAS meeting, the HZT already had similar, tentative evidence for acceleration in their own SN Ia data set. The HZT continued to perform numerous checks of their data analysis and interpretation, including fairly thorough consideration of various possible systematic effects. Unable to find any significant problems, even with the possible systematic effects, the HZT reported detection of a *nonzero* value for Ω_Λ (based on 16 high-z SNe Ia) at the Los Angeles dark matter conference in February 1998 (Filippenko & Riess 1998), and soon thereafter submitted a formal paper that was published in September 1998 (Riess et al. 1998b).

The HZT's original Hubble diagram for the 10 best-observed high-z SNe Ia is given in Figure 3. With the MLCS method applied to the full set of 16 SNe Ia, the HZT's formal results were $\Omega_M = 0.24 \pm 0.10$ if $\Omega_{\text{total}} = 1$, or $\Omega_M = -0.35 \pm 0.18$ (unphysical) if $\Omega_\Lambda = 0$. If one demanded that $\Omega_M = 0.2$, then the best value for Ω_Λ was 0.66 ± 0.21. These conclusions did not change significantly when only the 10 best-observed SNe Ia were used (Figure 3; $\Omega_M = 0.28 \pm 0.10$ if $\Omega_{\text{total}} = 1$).

Another important constraint on the cosmological parameters could be obtained from measurements of the angular scale of the first acoustic peak of the CMB (e.g., Zaldarriaga, Spergel, & Seljak 1997; Eisenstein, Hu, & Tegmark 1998); the SN Ia and CMB techniques provide nearly complementary information. A stunning result was already available by mid-1998 from existing measurements (e.g., Hancock et al. 1998; Lineweaver & Barbosa 1998): the HZT's analysis of the SN Ia data in Riess et al. (1998b) demonstrated that $\Omega_M + \Omega_\Lambda = 0.94 \pm 0.26$ (Figure 4), when the SN and CMB constraints were combined (Garnavich et al. 1998b; see also Lineweaver 1998, Efstathiou et al. 1999, and others).

Somewhat later (June 1999), the SCP published almost identical results (Perlmutter et al. 1999), implying an accelerating expansion of the Universe, based on an essentially independent set of 42 high-z SNe (although a substantial number of these were subsequently omitted because they are not clearly SNe Ia; Knop et al. 2003). The Perlmutter et al. (1999) data, together with those of the HZT, are shown in Figure 5, and the corresponding confidence contours in the Ω_Λ vs. Ω_M plane are given in Figure 6. This incredible agreement suggested that neither group had made a large, simple blunder; if the result was

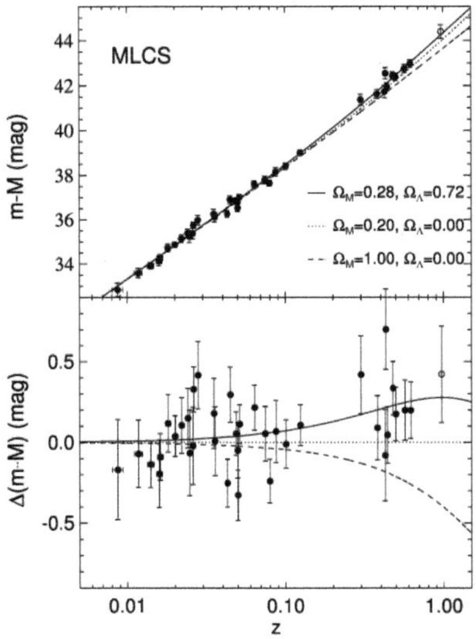

Figure 3. The upper panel shows the Hubble diagram for the low-z and high-z HZT SN Ia sample with MLCS distances; see Riess et al. (1998b). Overplotted are three world models: "low" and "high" Ω_M with $\Omega_\Lambda = 0$, and the best fit for a flat universe ($\Omega_M = 0.28$, $\Omega_\Lambda = 0.72$). The bottom panel shows the difference between data and models from the $\Omega_M = 0.20$, $\Omega_\Lambda = 0$ prediction. Only the 10 best-observed high-z SNe Ia are shown. The average difference between the data and the $\Omega_M = 0.20$, $\Omega_\Lambda = 0$ prediction is ~ 0.25 mag.

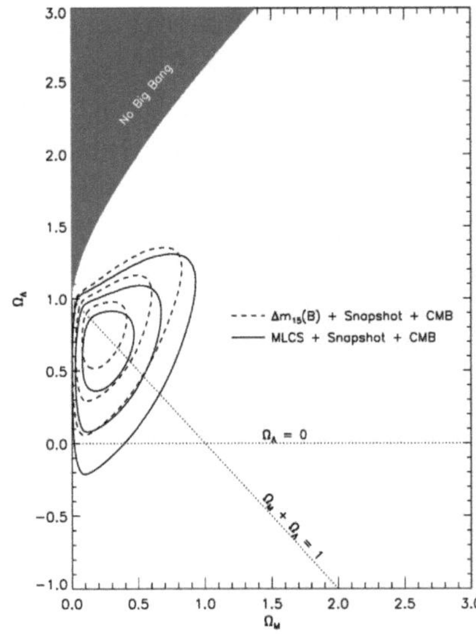

Figure 4. The HZT's combined constraints from SNe Ia (left) and the position of the first acoustic peak of the cosmic microwave background (CMB) angular power spectrum, based on data available in mid-1998; see Garnavich et al. (1998b). The contours mark the 68.3%, 95.4%, and 99.7% enclosed probability regions. Solid curves correspond to results from the MLCS method; dotted ones are from the $\Delta m_{15}(B)$ method; all 16 SNe Ia in Riess et al. (1998b) were used.

wrong, the reason must be subtle. Had there been only one team working in this area, it is likely that far fewer astronomers and physicists throughout the world would have taken the result seriously.

Moreover, already in 1998–1999 there was tentative evidence that the "dark energy" driving the accelerated expansion was indeed consistent with the cosmological constant, Λ. If Λ dominates, then the equation-of-state parameter of the dark energy should be $w = -1$, where the pressure (P) and density (ρ) are related according to $w = P/(\rho c^2)$. Garnavich et al. (1998b) and Perlmutter et al. (1999) were able to set an interesting limit, $w \lesssim -0.60$ at the 95% confidence level. However, more high-quality data at $z \approx 0.5$ are needed to narrow the allowed range, in order to test other proposed candidates for dark energy such as various forms of "quintessence" (e.g., Caldwell, Davé, & Steinhardt 1998).

Although the CMB results appeared reasonably persuasive in 1998–1999, one could argue that fluctuations on different scales had been measured with different instruments, and that subtle systematic effects might lead to erroneous conclusions. These fears were dispelled only 1–2 years later, when the more accurate and precise results of the BOOMERANG collaboration were announced (de Bernardis et al. 2000, 2002). Shortly thereafter the MAXIMA collaboration distributed their very similar findings (Hanany et al. 2000; Balbi et al. 2000; Netterfield et al. 2002; see also the TOCO, DASI, and many other measurements). Figure 6 illustrates that the CMB measurements tightly constrain $\Omega_{\rm total}$ to be close to unity; we appear to live in a flat universe, in agreement with most inflationary models for the early Universe! Combined with the SN Ia results, the evidence for nonzero Ω_Λ was fairly strong. Making the argument even more compelling was the fact that various studies of clusters of galaxies (see summary by Bahcall et al. 1999) showed that $\Omega_M \approx 0.3$, consistent with the results in Figures 4 and 6. Thus, a "concordance cosmology" had emerged: $\Omega_M \approx 0.3$, $\Omega_\Lambda \approx 0.7$ — consistent with what had been suspected some years earlier by Ostriker & Steinhardt (1995; see also Carroll, Press, & Turner 1992).

Yet another piece of evidence for a nonzero value of Λ was provided by the Two-Degree Field Galaxy Redshift Survey (2dFGRS; Peacock et al. 2001; Percival et al. 2001; Efstathiou et al. 2002). Combined with the CMB maps, their results are inconsistent with a universe dominated by gravitating dark matter. Again, the implication is that about 70% of the mass-energy density of the Universe consists of some sort of dark energy whose gravitational effect is repulsive. In 2003, results from the first year of *Wilkinson Microwave Anisotropy Probe (WMAP)* observations appeared; together with the 2dFGRS constraints, they confirmed and refined the concordance cosmology ($\Omega_M = 0.27$, $\Omega_\Lambda = 0.73$, $\Omega_{\rm baryon} = 0.044$, $H_0 = 71 \pm 4$ km s^{-1} Mpc^{-1}; Spergel et al. 2003, and references therein). Recent evidence for dark energy and an accelerating universe has also come from 2–3σ detections of the integrated Sachs-Wolfe effect (Afshordi, Loh, & Strauss 2004; Boughn & Crittenden 2004; Fosalba et al. 2003; Nolta et al. 2004; Scranton et al. 2004).

Meanwhile, the SN Ia measurements were becoming more numerous and of higher quality. For the HZT, the new (Fall 1999) sample of high-redshift SNe Ia presented by Tonry et al. (2003), analyzed with methods distinct from (but similar to) those used previously, confirmed the result of Riess et al. (1998b) and Perlmutter et al. (1999) that the expansion of the Universe is accelerating. By combining *all* of the data sets available at that time, Tonry et al. (2003) were able to use about 200 SNe Ia, obtaining an incredibly firm detection of $\Omega_\Lambda > 0$. They placed the following constraints on cosmological quantities:

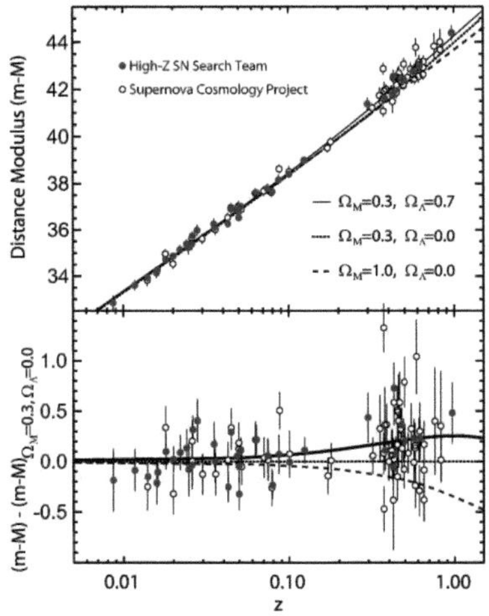

Figure 5. As in Figure 3, but this time including both the HZT (Riess et al. 1998b) and SCP (Perlmutter et al. 1999) samples of low-redshift and high-redshift SNe Ia. Overplotted are three world models: $\Omega_M = 0.3$ and 1.0 with $\Omega_\Lambda = 0$, and a flat universe ($\Omega_{\rm total} = 1.0$) with $\Omega_\Lambda = 0.7$. The bottom panel shows the difference between data and models from the $\Omega_M = 0.3$, $\Omega_\Lambda = 0$ prediction.

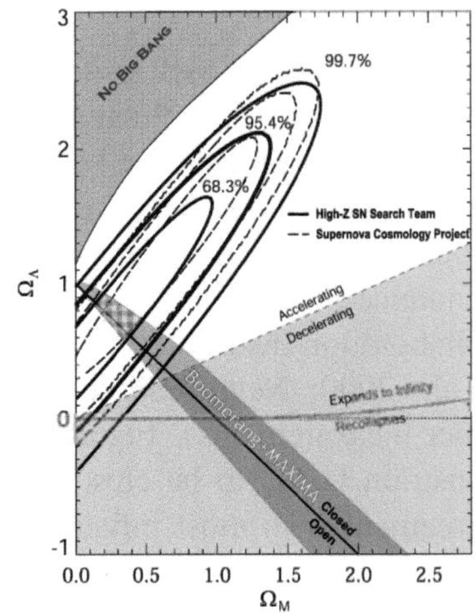

Figure 6. The combined constraints from SNe Ia (see Figure 5) and the position of the first acoustic peak of the CMB angular power spectrum, based on BOOMERANG and MAXIMA data. The contours mark the 68.3%, 95.4%, and 99.7% enclosed probability regions determined from the SNe Ia. According to the CMB, $\Omega_{\rm total} \approx 1.0$.

(1) If the equation-of-state parameter of the dark energy is $w = -1$, then $H_0 t_0 = 0.96 \pm 0.04$, and $\Omega_\Lambda - 1.4\Omega_M = 0.35 \pm 0.14$. (2) Including the constraint of a flat universe, they find that $\Omega_M = 0.28 \pm 0.05$, independent of any large-scale structure measurements. (3) Adopting a prior based on the 2dFGRS constraint on Ω_M (Percival et al. 2001) and assuming a flat universe, they derive that $-1.48 < w < -0.72$ at 95% confidence. (4) Adopting the 2dFGRS results, they find $\Omega_M = 0.28$ and $\Omega_\Lambda = 0.72$, independent of any assumptions about $\Omega_{\rm total}$. These constraints are similar in precision and in value to conclusions reported using Year 1 of *WMAP* (Spergel et al. 2003), also in combination with the 2dFGRS.

Even more recently, Barris et al. (2004) describe results from the 2001 HZT campaign. 23 SNe with $z = 0.34$–1.03 were studied, 9 of which are unambiguously SNe Ia. The sample includes 15 objects at $z > 0.7$, doubling the number of published SNe at these redshifts. Under the assumption that $\Omega_{\rm total} = 1$, they determine best-fit values of (Ω_M, Ω_Λ) = (0.33, 0.67), using 22 SNe from the 2001 HZT survey together with 172 SNe from Tonry et al. (2003) satisfying $z > 0.01$ and $A_V \leq 0.5$ mag.

The SCP's high-quality *HST* data set on 11 SNe Ia in the redshift range 0.36–0.86, recently published by Knop et al. (2003), independently confirms the apparent acceleration of the Universe. They were able to measure accurate colors of the SNe, providing better host-galaxy extinction estimates than had been possible in the past for individual objects. Thus, there was no need to make any assumptions or priors on the parent distribution of extinction values, $E(B-V)$, unlike the case in the initial analyses (Riess et al. 1998; Perlmutter et al. 1999; but see Sullivan et al. 2003, discussed below). Their extinction measurements do not show evidence for anomalously blue SNe Ia at high redshifts (or for a preponderance of negative $E(B-V)$ values for high-redshift SNe Ia), contrary to some earlier suspicions (Falco et al. 1999; Leibundgut 2001). Knop et al. (2003) find that dark energy is required with a probability exceeding 99%, consistent with previous studies that had made assumptions about the distribution of extinctions or that had used low-extinction subsets of SNe Ia. In a flat universe with a constant dark energy equation-of-state parameter $w=-1$, they find that $\Omega_\Lambda = 0.75\ (+0.06, -0.07) \pm 0.04$, where the first quoted uncertainties are statistical and the second are identified systematics. Moreover, in a flat universe with w constant in time, their SNe Ia data show that $w = -1.05(+0.15, -0.20) \pm 0.09$, consistent with Λ (rather than with some quintessence models; see below).

The dynamical age of the Universe can be calculated from the cosmological parameters. In an empty Universe with no cosmological constant, the dynamical age is simply the inverse of the Hubble constant, $t_0 = H_0^{-1}$; there is no deceleration. In the late-1990s, SNe Ia gave $H_0 = 65 \pm 7$ km s^{-1} Mpc^{-1}, and a Hubble time of 15.1 ± 1.6 Gyr. For a more complex cosmology, integrating the velocity of the expansion from the current epoch ($z=0$) to the beginning ($z=\infty$) yields an expression for the dynamical age. As shown in detail by Riess et al. (1998b), by mid-1998 the HZT had obtained a value of $14.2^{+1.0}_{-0.8}$ Gyr (with $H_0 = 65$) using the likely range for $(\Omega_M, \Omega_\Lambda)$ that they measured. (The precision was so high because their experiment was sensitive to roughly the *difference* between Ω_M and Ω_Λ, and the dynamical age also varies in approximately this way.) Including the *systematic* uncertainty of the Cepheid distance scale, which may be up to 10%, a reasonable estimate of the dynamical age was 14.2 ± 1.7 Gyr (Riess et al. 1998b). Again, the SCP's result was very similar (Perlmutter et al. 1999), since it was based on nearly the same derived values for the cosmological parameters. The most recent results, reported by Tonry et al. (2003; see also Knop et al. 2003) and adopting $H_0 = 72 \pm 8$ km s^{-1} Mpc^{-1}, give a dynamical age of 13.1 ± 1.5 Gyr for the Universe — again, in agreement with the *WMAP* result of 13.7 ± 0.2 Gyr.

This expansion age is also consistent with ages determined from various other techniques such as the cooling of white dwarfs (Galactic disk > 9.5 Gyr; Oswalt et al. 1996), radioactive dating of stars via the thorium and europium abundances (15.2 ± 3.7 Gyr; Cowan et al. 1997), and studies of globular clusters (10–15 Gyr, depending on whether *Hipparcos* parallaxes of Cepheids are adopted; Gratton et al. 1997; Chaboyer et al. 1998). The ages of the oldest stars no longer seem to exceed the expansion age of the Universe; the long-standing "age crisis" has evidently been resolved.

5. Discussion

Although the convergence of different methods on the same answer is reassuring, and suggests that the concordance cosmology is correct, it is important to vigorously test each method to make sure it is not leading us astray. Moreover, only through such detailed studies will the accuracy and precision of the methods improve, allowing us to eventually set better constraints on the equation-of-state parameter, w. Here I discuss the systematic effects that could adversely affect the SN Ia results.

High-redshift SNe Ia are observed to be dimmer than expected in an empty Universe (i.e., $\Omega_M = 0$) with no cosmological constant. At $z \approx 0.5$, where the SN Ia observations have their greatest leverage on Λ, the difference between an $\Omega_M = 0.3$ ($\Omega_\Lambda = 0$) universe and a flat universe with $\Omega_\Lambda = 0.7$ is only about 0.25 mag. Thus, we need to find out if chemical abundances, stellar populations, selection bias, gravitational lensing, or grey dust can have an effect this large. Although both the HZT and SCP had considered many of these potential systematic effects in their original discovery papers (Riess et al. 1998b; Perlmutter et al. 1999), and had shown with reasonable confidence that obvious ones were not greatly affecting their conclusions, it was of course possible that they were wrong, and that the data were being misinterpreted.

Gravitational Lensing

The magnification and demagnification of light by large-scale structure can alter the observed magnitudes of high-redshift SNe (e.g., Kantowski, Vaughan, & Branch 1995; Kantowski 1998). The effect of weak gravitational lensing on our analysis has been quantified by Wambsganss, Cen, & Ostriker (1998) and summarized by Schmidt et al. (1998). SN Ia light will, on average, be demagnified by 0.5% at $z = 0.5$ and by 1% at $z = 1$ in a Universe with a non-negligible cosmological constant. Although the sign of the effect is the same as the influence of a cosmological constant, the size of the effect is small enough to be ignored.

Holz (1998), Holz & Wald (1998), and Kantowski (1998) have calculated the weak lensing effects on supernova light from ordinary matter which is not smoothly distributed in galaxies but rather clumped into stars (i.e., dark matter contained in massive compact halo objects). With this scenario, microlensing becomes a more important effect, further decreasing the typical observed supernova luminosities at $z = 0.5$ by 0.02 mag for $\Omega_M = 0.2$. Even if most ordinary matter were contained in compact objects, this effect would not be large enough to reconcile the SN Ia distances with the influence of ordinary matter alone in a decelerating universe. Barris et al. (2004), for example, show that use of the "empty-beam model" for gravitational lensing does not eliminate the need for $\Omega_\Lambda > 0$. However, gravitational lensing will certainly need to be taken into account when making a precise measurement of w.

With a very large sample (200–1000) of high-redshift SNe Ia, and accurate photometry, it should be possible to quantify the effects of clumped matter. Light from some SNe Ia should be strongly amplified by the presence of intervening matter, while the vast majority will be deamplified (e.g., Barber et al. 2000). The distribution of amplification factors can be used to determine the type of dark matter most prevalent in the Universe (compact objects, or smoothly distributed).

Evolution

Perhaps the most obvious possible systematic effect is *evolution* of SNe Ia over cosmic time, due to changes in metallicity, progenitor mass, or some other factor. If the peak luminosity of SNe Ia were lower at high redshift, then the case for $\Omega_\Lambda > 0$ would weaken. Conversely, if the distant explosions are more powerful, then the case for acceleration strengthens. Theorists are not yet sure what the sign of the effect will be, if it is present at all; different assumptions lead to different conclusions (Höflich, Wheeler, & Thielemann 1998; Umeda et al. 1999; Nomoto et al. 2000; Yungelson & Livio 2000).

Of course, it is extremely difficult, if not effectively impossible, to obtain an accurate, independent measure of the peak luminosity of high-z SNe Ia, and hence to directly test for luminosity evolution. However, we can more easily determine whether *other* observable properties of low-z and high-z SNe Ia differ. If they are all the same, it is more probable that the peak luminosity is constant as well — but if they differ, then the peak luminosity might also be affected (e.g., Höflich et al. 1998). Drell, Loredo, & Wasserman (2000), for example, argue that there are reasons to suspect evolution, because the average properties of existing samples of high-z and low-z SNe Ia seem to differ (e.g., the high-z SNe Ia are more uniform).

The local sample of SNe Ia displays a weak correlation between light-curve shape and host-galaxy type, in the sense that the most luminous SNe Ia with the broadest light curves only occur in late-type galaxies. Both early-type and late-type galaxies provide hosts for dimmer SNe Ia with narrower light curves (Hamuy et al. 1996a). The mean luminosity difference for SNe Ia in late-type and early-type galaxies is ~ 0.3 mag. In addition, the SN Ia rate per unit luminosity is almost twice as high in late-type galaxies as in early-type galaxies at the present epoch (Cappellaro et al. 1997). These results may indicate an evolution of SNe Ia with progenitor age. Possibly relevant physical parameters are the mass, metallicity, and C/O ratio of the progenitor (Höflich et al. 1998).

We expect that the relation between light-curve shape and peak luminosity that applies to the range of stellar populations and progenitor ages encountered in the late-type and early-type hosts in our nearby sample should also be applicable to the range encountered in our distant sample. In fact, the age range for SN Ia progenitors in the nearby sample is likely *larger* than the change in mean progenitor age over the 4–6 Gyr lookback time to the high-z sample. Thus, to first order at least, our local sample should correct the distances for progenitor or age effects.

We can place empirical constraints on the effect that a change in the progenitor age would have on our SN Ia distances by comparing subsamples of low-redshift SNe Ia believed to arise from old and young progenitors. In the nearby sample, the mean difference between the distances for the early-type hosts (8 SNe Ia) and late-type hosts (19 SNe Ia), at a given redshift, is 0.04 $\pm$ 0.07 mag from the MLCS method. This difference is consistent with zero. Even if the SN Ia progenitors evolved from one population at low redshift to the other at high redshift, we still would not explain the surplus in mean distance of 0.25 mag over the $\Omega_\Lambda = 0$ prediction.

Moreover, in a major study of high-redshift SNe Ia as a function of galaxy morphology, the SCP found no clear differences (except for the amount of scatter; see §5.2) between the cosmological results obtained with SNe Ia in late-type and early-type galaxies (Sullivan et al. 2003). Similarly, a study of the host galaxies of high-redshift SNe Ia ($0.42 < z < 1.06$) done by the HZT (Williams et al. 2003) found no clear evidence for correlations between host-galaxy properties and the *residuals* of distance measurements from cosmological fits. Some of the correlations between SN Ia properties and host-galaxy type seen at low redshift also appear to be present at high redshift, again supporting the current practice of extrapolating properties of the nearby SN Ia population to high redshifts.

It is also reassuring that initial comparisons suggest that high-z SN Ia spectra appear similar to those observed at low redshift. For example, the spectral characteristics of SN 1998ai ($z = 0.49$) appear to be essentially indistinguishable from those of normal low-z SNe Ia; see Figure 7. In fact, the most obviously discrepant spectrum in this figure is the second one, that of SN 1994B ($z = 0.09$); it is intentionally included as a “decoy” that illustrates the degree to which even the spectra of nearby, relatively normal SNe Ia can vary. Nevertheless, it is important to note that a dispersion in luminosity (perhaps 0.2 mag) exists even among the other, more normal SNe Ia shown in Figure 7; thus, our spectra of SN 1998ai and other high-z SNe Ia are not yet sufficiently good for independent, *precise* determinations of peak luminosity from spectral features

(Nugent et al. 1995). Many of them, however, are sufficient for ruling out other SN types (Figure 8), or for identifying gross peculiarities such as those shown by SNe 1991T and 1991bg; see Coil et al. (2000).

We can help verify that the SNe at $z \approx 0.5$ being used for cosmology do not belong to a subluminous population of SNe Ia by examining rest-frame I-band light curves. Normal, nearby SNe Ia show a pronounced second maximum in the I band about a month after the first maximum and typically about 0.5 mag fainter (e.g., Ford et al. 1993; Suntzeff 1996). Subluminous SNe Ia, in contrast, do not show this second maximum, but rather follow a linear decline or show a muted second maximum (Filippenko et al. 1992a). As discussed by Riess et al. (2000), tentative evidence for the second maximum is seen from the HZT's existing J-band (rest-frame I-band) data on SN 1999Q ($z = 0.46$); see Figure 10. Additional tests with spectra and near-infrared light curves are currently being conducted.

Another way of using light curves to test for possible evolution of SNe Ia is to see whether the rise time (from explosion to maximum brightness) is the same for high-redshift and low-redshift SNe Ia; a difference might indicate that the peak luminosities are also different (Höflich et al. 1998). Riess et al. (1999c) measured the risetime of nearby SNe Ia, using data from KAIT, the Beijing Astronomical Observatory (BAO) SN search, and a few amateur astronomers. Though the exact value of the risetime is a function of peak luminosity, for typical low-redshift SNe Ia it is 20.0 ± 0.2 days. Riess et al. (1999b) pointed out that this differs by 5.8σ from the *preliminary* risetime of 17.5 ± 0.4 days reported in conferences by the SCP (Goldhaber et al. 1998a,b; Groom 1998). However, more thorough analyses of the SCP data (Aldering, Knop, & Nugent 2000; Goldhaber et al. 2001) show that the high-redshift uncertainty of ± 0.4 days that the SCP originally reported was much too small because it did not account for systematic effects. The revised discrepancy with the low-redshift risetime is about 2σ or less. Thus, the apparent difference in risetimes might be insignificant. Even if the difference is real, however, its relevance to the peak luminosity is unclear; the light curves may differ only in the first few days after the explosion, and this could be caused by small variations in conditions near the outer part of the exploding white dwarf that are inconsequential at the peak.

Although there are no clear signs that cosmic evolution of SNe Ia seriously compromises our results, it is wise to remain vigilant for possible problems. At low redshifts, for example, we already know that *some* SNe Ia don't conform with the correlation between light curve shape and luminosity. SN 2000cx in the S0 galaxy NGC 524, for example, has light curves that cannot be fit well by any of the fitting techniques currently available (Li et al. 2001a; Filippenko 2003); see Figure 9. Its late-time color is remarkably blue, inconsistent with the homogeneity described by Phillips et al. (1999). The spectral evolution

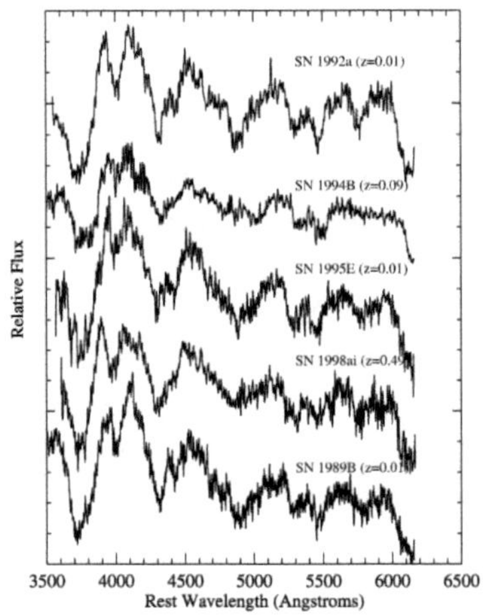

Figure 7. Spectral comparison (in f_λ) of SN 1998ai ($z = 0.49$; Keck spectrum) with low-redshift ($z < 0.1$) SNe Ia at a similar age ($\sim$ 5 days before maximum brightness), from Riess et al. (1998b). The spectra of the low-redshift SNe Ia were resampled and convolved with Gaussian noise to match the quality of the spectrum of SN 1998ai. Overall, the agreement in the spectra is excellent, tentatively suggesting that distant SNe Ia are physically similar to nearby SNe Ia. SN 1994B ($z = 0.09$) differs the most from the others, and was included as a "decoy."

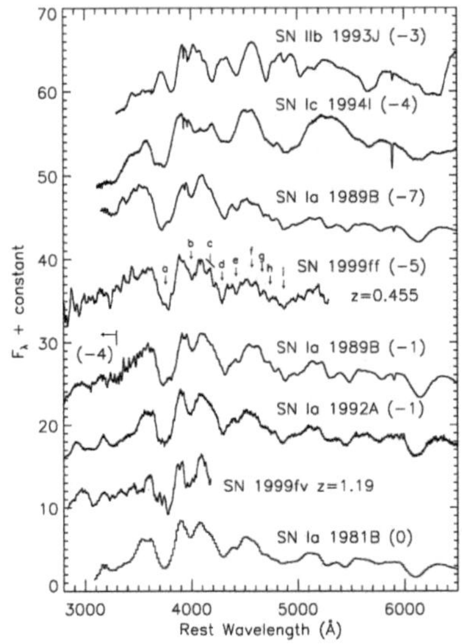

Figure 8. Heavily smoothed spectra of two high-z SNe (SN 1999ff at $z = 0.455$ and SN 1999fv at $z = 1.19$; quite noisy below $\sim$3500 Å) are presented along with several low-z SN Ia spectra (SNe 1989B, 1992A, and 1981B), a SN Ib spectrum (SN 1993J), and a SN Ic spectrum (SN 1994I); see Filippenko (1997b) for a discussion of spectra of various types of SNe. The date of the spectra relative to B-band maximum is shown in parentheses after each object's name. Specific features seen in SN 1999ff and labeled with a letter are discussed by Coil et al. (2000). This comparison shows that the two high-z SNe are most likely SNe Ia.

of SN 2000cx is peculiar as well (Li et al. 2001a; Branch et al. 2004a): the photosphere appears to have remained hot for a long time, and both iron-peak and intermediate-mass elements move at very high velocities.

An even *more* peculiar object is SN 2002cx (Li et al. 2003; Filippenko 2003; Branch et al. 2004b). It is spectroscopically bizarre, with extremely low expansion velocities and almost no evidence for intermediate-mass elements. The nebular phase was reached incredibly soon after maximum brightness, despite the low velocity of the ejecta, suggesting that the ejected mass is small. SN 2002cx was subluminous by $\sim$ 2 mag at all optical wavelengths relative to normal SNe Ia, despite the early-time spectroscopic resemblance to the some-

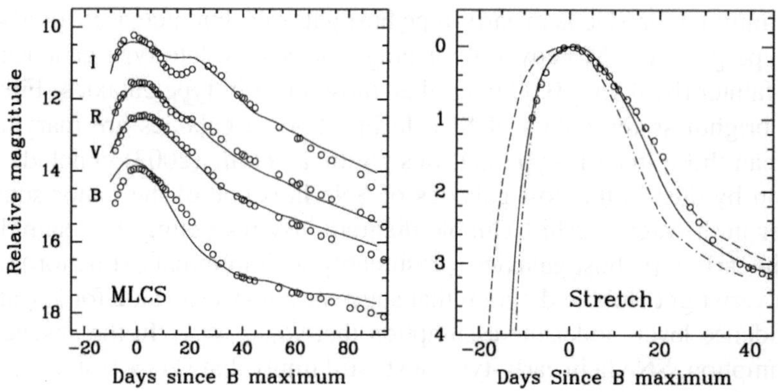

Figure 9. The MLCS fit (Riess et al. 1998b; left panel) and the stretch method fit (Perlmutter et al. 1999; right panel) for SN 2000cx. The MLCS fit is the worst we had ever seen through year 2000. For the stretch method fit, the solid line is the fit to all the data points from $t = -8$ to 32 days, the dash-dotted line uses only the premaximum datapoints, and the dashed line only the postmaximum datapoints. The three fits give very different stretch factors. From Li et al. (2001a).

what overluminous SN 1991T. The R-band and I-band light curves of SN 2002cx are completely unlike those of normal SNe Ia. No existing theoretical model successfully explains all observed aspects of SN 2002cx, though 3D deflagration models may be best. If there are more strange beasts like SNe 2000cx and 2002cx at high redshifts than at low redshifts, systematic errors may creep into the analysis of high-z SNe Ia.

Extinction

Our SN Ia distances have the important advantage of including corrections for interstellar extinction occurring in the host galaxy and the Milky Way. Extinction corrections based on the relation between SN Ia colors and luminosity improve distance precision for a sample of nearby SNe Ia that includes objects with substantial extinction (Riess et al. 1996a; Phillips et al. 1999); the scatter in the Hubble diagram is much reduced. Moreover, the consistency of the measured Hubble flow from SNe Ia with late-type and early-type hosts (see §5.1) shows that the extinction corrections applied to dusty SNe Ia at low redshift do not alter the expansion rate from its value measured from SNe Ia in low-dust environments.

In practice, the high-redshift SNe Ia generally appear to suffer very little extinction; their $B - V$ colors at maximum brightness are normal, suggesting little color excess due to reddening. The most detailed available study is that of the SCP (Sullivan et al. 2003): they found that the scatter in the Hubble diagram is minimal for SNe Ia in early-type host galaxies, but increases for SNe Ia in late-type galaxies. Moreover, on average the SNe in late-type galaxies are slightly fainter (by 0.14 ± 0.09 mag) than those in early-type galaxies. Finally, at peak brightness the colors of SNe Ia in late-type galaxies are marginally redder than those in early-type galaxies. Sullivan et al. (2003) conclude that extinction by dust in the host galaxies of SNe Ia is one of the major sources of scatter in the high-redshift Hubble diagram. By restricting their sample to SNe Ia in early-type host galaxies (presumably with minimal extinction), they obtain a very tight Hubble diagram that suggests a nonzero value for Ω_Λ at the 5σ confidence level, under the assumption that $\Omega_{\rm total} = 1$. In the absence of this assumption, SNe Ia in early-type hosts still imply that $\Omega_\Lambda > 0$ at nearly the 98% confidence level. The results for Ω_Λ with SNe Ia in late-type galaxies are quantitatively similar, but statistically less secure because of the larger scatter.

Riess, Press, & Kirshner (1996b; see also Phillips et al. 1999) found indications that the Galactic ratios between selective absorption and color excess are similar for host galaxies in the nearby ($z \leq 0.1$) Hubble flow. Yet, what if these ratios changed with lookback time (e.g., Aguirre 1999a)? Could an evolution in dust-grain size descending from ancestral interstellar "pebbles" at higher redshifts cause us to underestimate the extinction? Large grains would not imprint the reddening signature of typical interstellar extinction upon which our corrections necessarily rely.

However, viewing our SNe through such gray interstellar grains would also induce a *dispersion* in the derived distances. Using the results of Hatano, Branch, & Deaton (1998), Riess et al. (1998b) estimate that the expected dispersion would be 0.40 mag if the mean gray extinction were 0.25 mag (the value required to explain the measured MLCS distances without a cosmological constant). This is significantly larger than the 0.21 mag dispersion observed in the high-redshift MLCS distances. Furthermore, most of the observed scatter is already consistent with the estimated *statistical* errors, leaving little to be caused by gray extinction. Nevertheless, if we assumed that *all* of the observed scatter were due to gray extinction, the mean shift in the SN Ia distances would be only 0.05 mag. With the existing observations, it is difficult to rule out this modest amount of gray interstellar extinction.

Gray *intergalactic* extinction could dim the SNe without either telltale reddening or dispersion, if all lines of sight to a given redshift had a similar column density of absorbing material. The component of the intergalactic medium with such uniform coverage corresponds to the gas clouds producing Lyman-α forest absorption at low redshifts. These clouds have individual H I column densities less than about $10^{15}\ {\rm cm}^{-2}$ (Bahcall et al. 1996). However, they display low metallicities, typically less than 10% of solar. Gray extinction would require larger dust grains which would need a larger mass in heavy elements than typical interstellar grain size distributions to achieve a given extinction. It

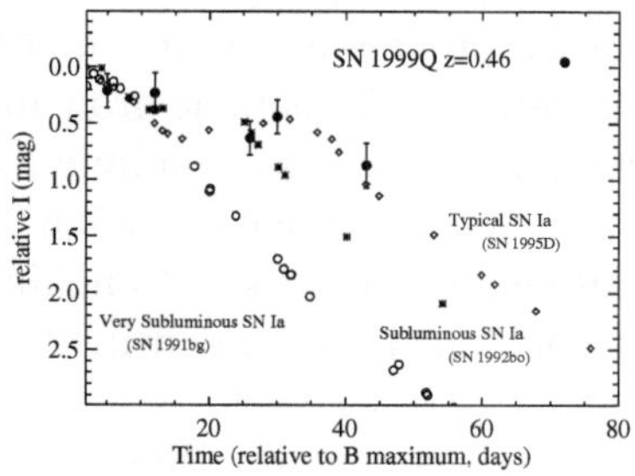

Figure 10. Rest-frame I-band (observed J-band) light curve of SN 1999Q ($z = 0.46$, 5 solid points; Riess et al. 2000), and the I-band light curves of several nearby SNe Ia. Subluminous SNe Ia exhibit a less prominent second maximum than do normal SNe Ia.

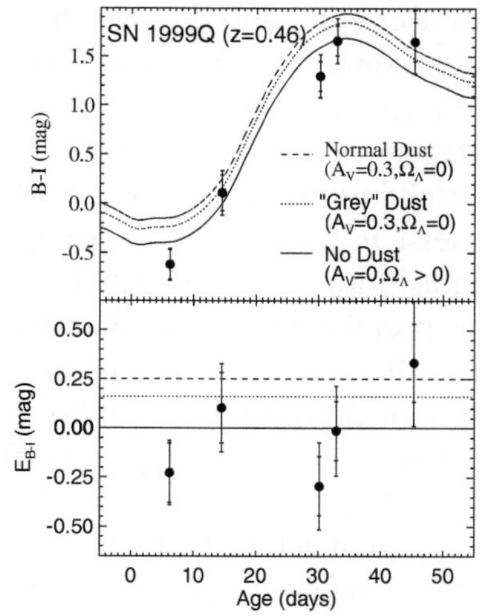

Figure 11. Color ($B - I$) and color excess (E_{B-I}) for SN 1999Q and different dust models (Riess et al. 2000). The data are most consistent with no dust and $\Omega_\Lambda > 0$.

is possible that large dust grains are blown out of galaxies by radiation pressure, and are therefore not associated with Lyman-α clouds (Aguirre 1999b).

But even the dust postulated by Aguirre (1999a,b) and Aguirre & Haiman (1999) is not *completely* gray, having a size of about 0.1 μm. We can test for such nearly gray dust by observing high-redshift SNe Ia over a wide wavelength range to measure the color excess it would introduce. If $A_V = 0.25$ mag, then $E(U-I)$ and $E(B-I)$ should be 0.12–0.16 mag (Aguirre 1999a,b). If, on the other hand, the 0.25 mag faintness is due to Λ, then no such reddening should be seen. This effect is measurable using proven techniques; so far, with just one SN Ia (SN 1999Q; Figure 11), our results favor the no-dust hypothesis to better than 2σ (Riess et al. 2000). More work along these lines is in progress.

Early Deceleration of the Universe

Suppose, however, that for some reason the dust is *very* gray, or our color measurements are not sufficiently precise to rule out Aguirre's (or other) dust. Or, perhaps some other astrophysical systematic effect is fooling us, such as possible evolution of the white dwarf progenitors (e.g., Höflich et al. 1998; Umeda et al. 1999), or gravitational lensing (Wambsganss et al. 1998). The most decisive test to distinguish between Λ and cumulative systematic effects is to examine the *deviation* of the observed peak magnitude of SNe Ia from the magnitude expected in the low-Ω_M, zero-Λ model. If Λ is positive, the deviation should actually begin to *decrease* at $z \approx 1$; we will be looking so far back in time that the Λ effect becomes small compared with Ω_M, and the Universe is *decelerating* at that epoch. If, on the other hand, a systematic bias such as gray dust or evolution of the white dwarf progenitors is the culprit, we expect that the deviation of the apparent magnitude will continue growing, unless the systematic bias is set up in such an unlikely way as to mimic the effects of Λ (Drell et al. 2000). A turnover, or decrease of the deviation of apparent magnitude at high redshift, can be considered almost like a "smoking gun" of Λ (or, more generally, of a dark-energy component whose density does not change much with time).

In a wonderful demonstration of good luck and hard work, Riess et al. (2001) report on *HST* observations of a probable SN Ia at $z \approx 1.7$ (SN 1997ff, the most distant SN ever observed) that suggest the expected turnover is indeed present, providing a tantalizing glimpse of the epoch of deceleration. (See also Benítez et al. 2002, which corrects the observed magnitude of SN 1997ff for gravitational lensing by foreground galaxies.) SN 1997ff was discovered by Gilliland & Phillips (1998) in a repeat *HST* observation of the Hubble Deep Field–North, and it was serendipitously monitored in the infrared with *HST*/NICMOS. The peak apparent SN brightness is consistent with that expected in the decelerating phase of the concordance cosmological model, $\Omega_M \approx 0.3$, $\Omega_\Lambda \approx 0.7$ (Figure 12). It is inconsistent with gray dust or simple luminosity evolution, when combined with the data for SNe Ia at $z \approx 0.5$.

On the other hand, it was wise to remain cautious at the time: the error bars are large, and it is always possible that we are being fooled by this single object. The HZT and SCP thus started programs to find and measure more SNe Ia at such high redshifts ($z > 1$). One promising-looking result was that of Tonry et al. (2003) and Barris et al. (2004) for the HZT: the deviation of apparent magnitude from the low-Ω_M, zero-Λ model for several new SNe Ia at $z \approx 1$ is roughly the same as that at $z \approx 0.5$, in agreement with expectations based on the results of Riess et al. (2001).

Inspired by his promising results from SN 1997ff, as well as by the discovery and study of two distant SNe Ia (Blakeslee et al. 2003) with the *HST* Advanced Camera for Surveys (ACS), Adam Riess launched an extensive *HST* campaign (the "Hubble Higher-z Supernova Search") to closely monitor ~ 6 high-redshift ($z \gtrsim 1$) SNe Ia, with the goal of more clearly detecting the early epoch of deceleration. A major obstacle was the vast amount of *HST* time required to both discover and obtain follow-up photometry of the faint SNe, but this was overcome by arranging to "piggyback" on the Great Observatories Origins Deep Survey (GOODS), in which ~ 400 *HST* orbits were being

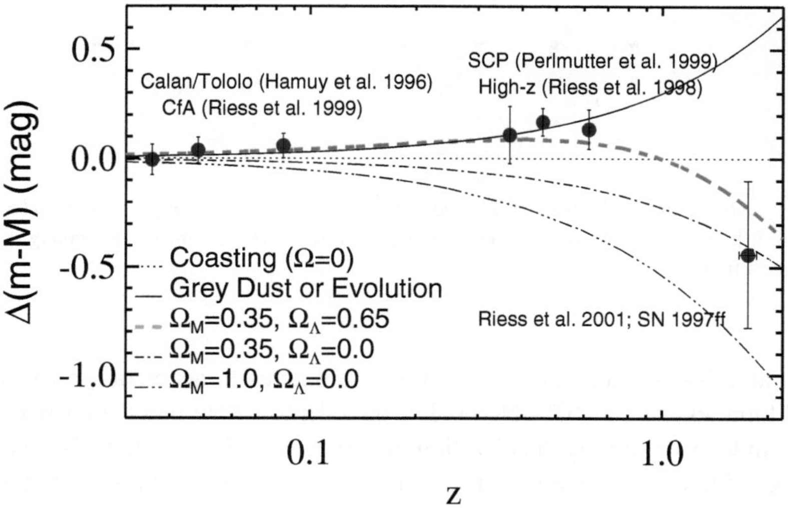

Figure 12. Hubble diagram for SNe Ia relative to an empty universe ($\Omega = 0$) compared with cosmological and astrophysical models (Riess et al. 2001). Low-redshift SNe Ia are from Hamuy et al. (1996a) and Riess et al. (1999a). The magnitude of SN 1997ff at $z = 1.7$ has been corrected for gravitational lensing (Benítez et al. 2002). The measurements of SN 1997ff are inconsistent with astrophysical effects that could mimic previous evidence for an accelerating universe from SNe Ia at $z \approx 0.5$.

used to obtain deep, multicolor ACS images of distant galaxies. The GOODS team obtained their data in several distinct epochs with separations of 45 days, and with suitable filters, allowing the discovery of high-redshift SNe (e.g., Giavalisco et al. 2002; Riess et al. 2004a), generally on the rise. Results from Jha (2002) suggested that key light-curve-shape parameters, such as the time of maximum and decline rate, could be determined from rest-frame U-band light curves, corresponding to observed-frame ACS z-band images, allowing for fewer epochs of costly infrared (NICMOS) observations to measure the SN luminosity and color. This strategy allowed the requisite follow-up images (e.g., Figure 13) and ACS grism spectra to be obtained with a relatively modest award of additional *HST* time.

The program was very successful. Riess et al. (2004b; see also Strolger et al. 2004) convincingly show that SNe Ia at $z \gtrsim 1$ are brighter than predicted in simple models invoking systematic effects (Fig. 14). Thus, many SN Ia

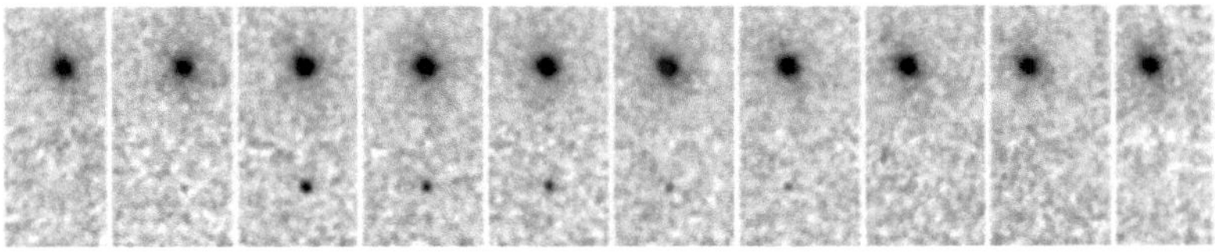

Figure 13. SN 2002hp, a high-redshift supernova from the GOODS program using *HST*. One can see it brightening and subsequently fading with time. The assumed host galaxy is at the top of each frame.

evolution and dust models are excluded as alternatives to acceleration for the observed faintness of $z \approx 0.5$ SNe Ia. The data demonstrate that the early universe was indeed apparently decelerating, as expected if Λ is a relatively recent effect (Fig. 15). The expansion of the Universe made a transition between deceleration and acceleration (Turner & Riess 2002) at $z \approx 0.5$, consistent with the new "concordance cosmology" of $\Omega_M \approx 0.27$, $\Omega_\Lambda \approx 0.73$ from the best CMB measurements (Spergel et al. 2003) and (independently) the complete set of available SN Ia data (Riess et al. 2004b). Finally, Riess et al. (2004b) show that the value of the dark energy equation-of-state parameter is consistent with $w = -1$, and that changes in w could not have been very large over the past $\sim$ 10 billion years (i.e., $dw/dz = w' \approx 0$, as expected with Λ). Even better agreement with the Λ-model expectations was obtained by Wang & Tegmark (2004), who analyzed the data in a slightly different manner. Thus, although we cannot exclude the possibility that the Universe will recollapse in the future, such a "big crunch" (or "gnaB giB," which is "Big Bang" backwards) is unlikely to occur in fewer than 15–30 Gyr (Riess et al. 2004b) if we adopt the linear potential field of Kallosh & Linde (2003).

Of course, it is possible to find specific dust or evolution models that are *not* ruled out with our SN Ia data. For example, a "replenishing dust" model represents a constant density of dust that is continually replenished at precisely the rate at which it is diluted by the expanding universe (Goobar, Bergstrom, & Mortsell 2002; Riess et al. 2004b); one could perhaps call this the "steady-state dust model." Since the dimming is directly proportional to the distance light traveled and is thus mathematically quite similar to the effects of a cosmological constant, we cannot discriminate this model from the Λ-dominated model in the magnitude-redshift plane. However, the fine-tuning required of this dust's opacity, replenishing rate, and velocity ($\gtrsim 1000$ km s^{-1} for it to fill space uniformly without adding detectable dispersion) makes it unattractive as a simpler alternative to a cosmological constant. Moreover, the density of intergalactic dust needed to explain the observed faintness of SNe Ia at $z \approx 0.5$ exceeds the upper limit determined by Paerels et al. (2002) from the absence of a detectable X-ray scattering halo around a single distant quasar.

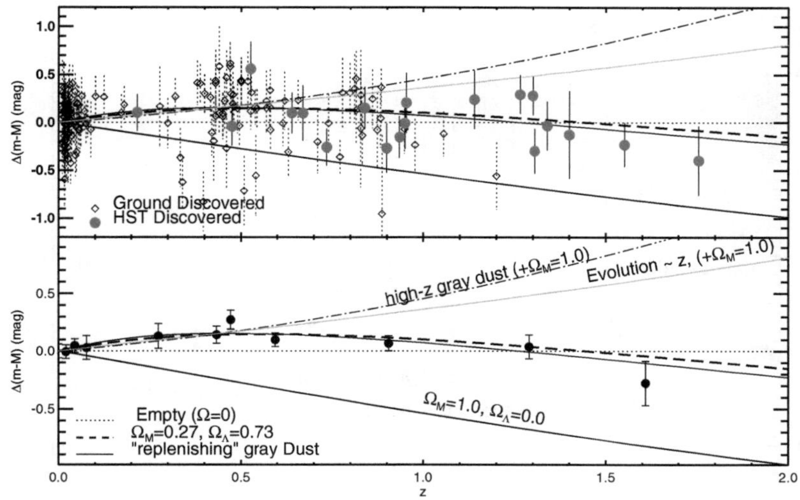

Figure 14. SN Ia residual Hubble diagram comparing cosmological models and scenarios for astrophysical dimming (Riess et al. 2004b). (Upper panel:) SNe Ia from ground-based discoveries are shown as diamonds, HST-discovered SNe Ia are shown as filled symbols. (Lower panel:) Weighted averages in fixed redshift bins, which are given only for illustrative purposes. Data and models are shown relative to an empty universe model ($\Omega = 0$).

Another model with this behavior is evolution that is proportional to look-back time (Wright 2001). While possible, such dimming behavior, especially if in the form of luminosity evolution, seems implausible. We may expect evolution (or dust production) to be coupled to the observed evolution of stellar populations, galaxies morphologies, sizes, large-scale structure, or even chemical enrichment. None of these known varieties of evolution are largely completed by $z = 0.5$ starting from their properties at $z = 0$; quite the contrary, most of them have hardly begun, looking back to $z = 0.5$. As mentioned previously, a strong empirical argument against recent luminosity evolution is the independence of SN Ia distance measurements on gross host morphology (Riess et al. 1998b; Sullivan et al. 2003). The range of progenitor formation environments spanned by SNe Ia in early-type and late-type hosts greatly exceeds any evolution in the mean host properties between $z = 0$ and $z = 0.5$. In the end, however, the only "proof" against astrophysical contamination of the cosmological signal from SNe Ia is to test the results against those of other observations, independent of SNe Ia.

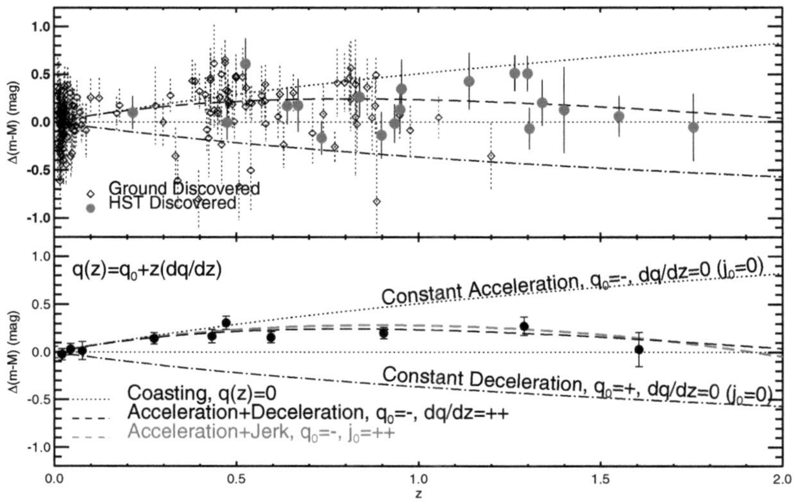

Figure 15. Kinematic SN Ia residual Hubble diagram (Riess et al. 2004b). (Upper panel:) Symbols as in Figure 14. (Lower panel:) Weighted averages, as in Fig. 14. Data and kinematic models of the expansion history are shown relative to an eternally coasting model, $q(z) = 0$. Models representing specific kinematic scenarios (e.g., "constant acceleration") are illustrated.

Type Ia Supernova Rates and Progenitors

The rate of SNe Ia as a function of redshift is important to know for a number of reasons. For example, it affects the derived chemical evolution of galaxies (e.g., Maoz & Gal-Yam 2004; Matteucci & Recchi 2001, and references therein), and it can be used to set constraints on the star formation history of the Universe (Gal-Yam & Maoz 2004). Moreover, the distribution of "delay times" between the formation of the progenitor star and the explosion of the supernova can be determined from an accurate census of the SN Ia rate at different cosmic times, thereby providing clues to the physical nature of the progenitors of SNe Ia (Madau, Della Valle, & Panagia 1998; Dahlen & Fransson 1999; Yungelson & Livio 2000). There have been many measurements of the rate of very nearby SNe Ia (Cappellaro, Evans, & Turatto 1999, and references therein), and at $z \approx 0.1$ several studies have been conducted (Hardin et al. 2000; Strolger 2003; Reiss 2000). These form the basis of comparison for the measurements at higher redshifts.

From observations of three SNe Ia at $z \approx 0.4$, Pain et al. (1996) estimated a rate of SNe Ia equivalent to 34 yr^{-1} deg^{-2}, with a 1σ uncertainty of a factor of ~ 2, for objects found in the range of $21.3 < R < 22.3$ mag. A more recent estimate by Pain et al. (2002a) is $(1.53 \pm 0.3) \times 10^{-4}\ h^3\ \mathrm{Mpc}^{-3}\ \mathrm{yr}^{-1}$ at a mean redshift of 0.55 (where $h \equiv H_0/(100\ \mathrm{km\ s}^{-1}\ \mathrm{Mpc}^{-1})$. Cappellaro et al. (1999) report a nearby SN Ia rate of $0.36 \pm 0.11\,h^2$ SNu (where 1 SNu $\equiv 10^{10}\ L_{B_\odot}$ per century), or $(0.79 \pm 0.24) \times 10^{-4}\ h^3\ \mathrm{Mpc}^{-3}\ \mathrm{yr}^{-1}$, using a quite uncertain local luminosity density of $\rho = 2.2 \times 10^8\ h\ L_{B_\odot}\ \mathrm{Mpc}^{-3}$. Pain et al. (2002a) claim to see a very modest increase in the rate of SNe Ia with redshift, perhaps tracking the star formation rate which Wilson et al. (2002) estimate as being proportional to $(1+z)^{1.7}$ in a flat universe with $\Omega_M = 0.3$.

Based on a new sample of SNe Ia observed during the 1999 HZT campaign, an independent estimate of the SN Ia rate at a mean redshift of 0.46 was made by Tonry et al. (2003): $(1.4 \pm 0.5) \times 10^{-4}\ h^3\ \mathrm{Mpc}^{-3}\ \mathrm{yr}^{-1}$. This is in excellent agreement with the results of Pain et al. (2002a), and it is not inconsistent with the local rates from Cappellaro et al. (1999), particularly given the uncertainty in the local luminosity density. On the other hand, from a different sample of SNe Ia (see below), Dahlen et al. (2004) find a somewhat higher rate of SNe Ia at $z \approx 0.4$ and suggest that the previously published rates may have been underestimated due to systematic effects. Indeed, B. Barris (2004, private communication) has found evidence that the true rate of SNe Ia in the 2001 HZT campaign is higher than that reported by Barris et al. (2004); similar biases may have affected the 1999 HZT campaign and other previous searches.

Although Tonry et al. (2003) were incomplete in their counts of SNe Ia at $z \approx 1$, they do not believe that the rate of SNe Ia closely tracks the star formation rate. Application of the star formation rate from Wilson et al. (2002) would suggest that the SN Ia rate at $z \approx 1.1$ should be three times as great as occurs locally and nearly twice as great as the rate at $z = 0.46$. In this case, Tonry et al. (2003) should have discovered 16 SNe Ia deg^{-2} in their search, yet they only found 4 SNe Ia deg^{-2}. It is their impression that the constant rate per volume is closer to the truth.

The more recent results of Dahlen et al. (2004) conflict with this conclusion. They determine the rate of SNe Ia as a function of redshift by using data from the Hubble Higher-z Supernova Search of Riess et al. (2004b), with a redshift range of $0.2 < z < 1.6$. The resulting SN Ia rate at $z \approx 1$ is a factor of 3–5 higher than previous estimates made at lower redshifts ($z < 0.5$), presumably because the star formation rate at $z > 1$ was substantially higher than that in the past few Gyr. Moreover, their data suggest that at even higher redshifts, $z > 1$, the rate of SNe Ia begins to decrease. They find that the delay time (from progenitor star formation to SN Ia explosion) is likely to be substantial, $\gtrsim 2$ Gyr. In addition, assuming a Salpeter (1955) initial mass function and a SN Ia progenitor main-sequence mass range of 3–8 $M_\odot$, it appears that 5–7% of the white dwarfs formed from these progenitor stars eventually become SNe Ia (but fewer than 1% of *all* white dwarfs eventually become SNe Ia).

Strolger et al. (2004) use the above results, together with additional modeling, to set more quantitative constraints on the nature of the progenitors of SNe Ia. Specifically, they use a Bayesian maximum likelihood test to determine the most likely range of delay times that best reproduces the observed redshift distribution of SNe Ia. They find that models requiring a large fraction of "prompt" (less than 2 Gyr) SNe Ia poorly reproduce the observed redshift distribution and are rejected at $> 99\%$ confidence. Thus, SNe Ia cannot generally be prompt events, nor can they be expected to closely follow the star formation rate history. Instead, Gaussian models best fit the observed data for mean delay times in the range of 3–4 Gyr. This may be most consistent with single-degenerate systems in which the white dwarf accretes from a main-sequence companion or from a somewhat evolved companion (Livio 2001), although certain types of double-degenerate models are not yet eliminated. Tests conducted by Gal-Yam & Maoz (2004) also conclude that the characteristic delay times of SNe Ia should be large (>1–2 Gyr) for similar assumed models of the star formation rate history, but the results are not as definitive as those of Strolger et al. (2004) because they are based on SN Ia rates derived from more limited SCP data (Pain et al. 2002a).

Measuring the Dark Energy Equation-of-State Parameter

Every energy component in the Universe can be parameterized by the way its density varies as the Universe expands (scale factor a), with $\rho \propto a^{-3(1+w)}$, and w is the component's equation-of-state parameter, $w = P/(\rho c^2)$, where P is the pressure exerted by the component. So for matter, $w = 0$, while an energy component that does not vary with scale factor has $w = -1$, as in the cosmological constant Λ. Quintessence models have $w \neq -1$, and generally $dw/dz \neq 0$). Some really strange energies may have $w < -1$: their density increases with time (Carroll, Hoffman, & Trodden 2003), leading to a "Big Rip" in which progressively smaller bound systems get torn apart! [Riess et al. (2004b) and Caldwell et al. (2003) estimate that such a fate will not occur sooner than $\sim$ 20 Gyr from now, if ever.] Clearly, a good estimate of w becomes the key to differentiating between models.

The CMB observations imply that the geometry of the universe is close to flat, so the energy density of the dark component is simply related to the matter density by $\Omega_x = 1 - \Omega_M$. This allows the luminosity distance as a function of redshift to be written as

$$D_L(z) = \frac{c(1+z)}{H_0} \int_0^z \frac{[1 + \Omega_x((1+\mathrm{z})^{3w} - 1)]^{-1/2}}{(1+\mathrm{z})^{3/2}} \, \mathrm{dz} \, ,$$

showing that the dark energy density and equation of state directly influence the apparent brightness of standard candles. As demonstrated graphically in

Figure 16, SNe Ia observed over a wide range of redshifts can constrain the dark energy parameters to a cosmologically interesting accuracy.

But there are two major problems with using SNe Ia to measure w. First, systematic uncertainties in SN Ia peak luminosity limit how well $D_L(z)$ can be measured. While statistical uncertainty can be arbitrarily reduced by finding thousands of SNe Ia, intrinsic SN properties such as evolution and progenitor metallicity, and observational limits like photometric calibrations and K-corrections, create a systematic floor that cannot be decreased by sheer force of numbers. We expect that systematics can be controlled to at best 3%, with considerable effort.

Second, SNe at $z > 1.0$ are very hard to discover and study from the ground. As discussed above, both the HZT and the SCP have found a few SNe Ia at $z > 1.0$, but the numbers and quality of these light curves are insufficient for a w measurement. Large numbers of SNe Ia at $z > 1.0$ are best left to a wide-field optical/infrared imager in space, such as the proposed *Supernova/ Acceleration Probe* (*SNAP*; Nugent et al. 2001) satellite.

Fortunately, an interesting measurement of w can be made at present. The current values of Ω_M from many methods (most recently *WMAP*: 0.27; Spergel et al. 2003) make an excellent substitute for those expensive SNe at $z > 1.0$. Figure 16 shows that a SN Ia sample with a maximum redshift of $z = 0.8$, combined with the current 10% error on Ω_M, will do as well as a SN Ia sample at much higher redshifts. Within a few years, the Sloan Digital Sky Survey and *WMAP* will solidify the estimate of Ω_M and sharpen w further.

Both the SCP and the HZT are involved in multi-year programs to discover and monitor hundreds of SNe Ia for the purpose of measuring w. For example, the HZT's project, ESSENCE (Equation of State: SupErNovae trace Cosmic Expansion), is designed to discover 200 SNe Ia evenly distributed in the $0.2 < z < 0.8$ range (Smith et al. 2002; Garnavich et al. 2002; http://www.ctio.noao.edu/wproject). The CTIO 4-m telescope and mosaic camera are being used to find and follow the SNe by imaging on every other dark night for several consecutive months of the year. Keck and other large telescopes are being used to get the SN spectra and redshifts. Project ESSENCE will eventually provide an estimate of w to an accuracy of $\sim$10% (Figure 17). Even larger numbers of high-redshift SNe Ia are being found during the ongoing CFHT Legacy Survey (http://www.cfht.hawaii.edu/Science/CFHLS; Pain et al. 2002b), providing an independent sample with which to measure the value of w. Within the next few years, Pan-STARRS (Kaiser et al. 2002; http://poi.ifa.hawaii.edu) should discover and follow thousands of SNe Ia.

Farther in the future, the plethora of SNe Ia to be found and studied by the proposed *SNAP* satellite (Nugent et al. 2001), the NASA/DOE Joint Dark Energy Mission (JDEM), the Large-area Synoptic Survey Telescope (the "Dark Matter Telescope"; Tyson & Angel 2001), and similar large-scale projects

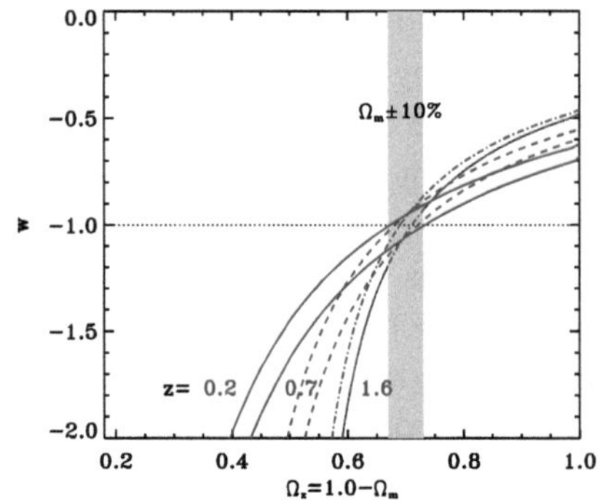

Figure 16. Constraints on Ω_x and w from SN data sets collected at $z = 0.2$ (solid lines), $z = 0.7$ (dashed lines), and $z = 1.6$ (dash-dot lines). The shaded area indicates how an independent estimate of Ω_M with a 10% error can help constrain w.

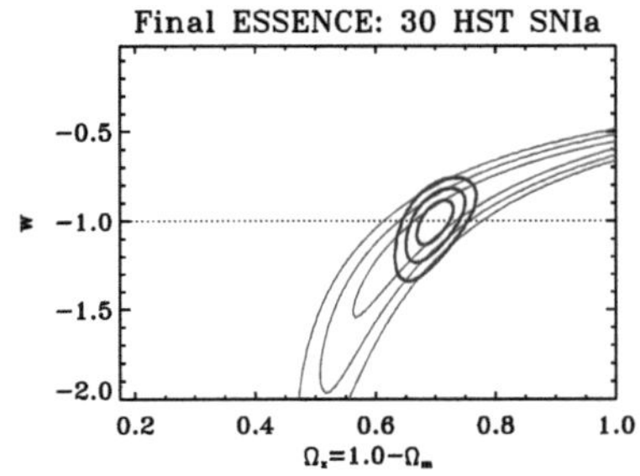

Figure 17. Expected constraints on w with the desired final ESSENCE data set of 200 SNe Ia, 30 of which (in the redshift range $0.6 < z < 0.8$) are to be observed with HST. The thin lines are for SNe alone while the thick lines assume an uncertainty in Ω_M of 7%. The final ESSENCE data will constrain the value of w to $\sim$10%.

could reveal whether the value of w depends on redshift, and hence should give additional constraints on the nature of the dark energy. High-redshift surveys of galaxies such as DEEP2 (Davis et al. 2001), as well as space-based missions to map the CMB (*Planck*), should provide additional evidence for (or against) Λ. Observational cosmology promises to remain exciting for quite some time!

Acknowledgments

I thank all of my HZT collaborators for their contributions to our team's research, and members of the SCP for their seminal complementary work on the accelerating Universe. My group's work at U.C. Berkeley has been supported by NSF grants AST–9987438, AST–0206329, and AST-0307894, as well as by grants GO–7505, GO/DD–7588, GO–8177, GO–8641, GO–9118, GO–9352, and GO–9728 from the Space Telescope Science Institute, which is operated by the Association of Universities for Research in Astronomy, Inc., under NASA contract NAS 5–26555. Many spectra of high-redshift SNe were obtained at the W. M. Keck Observatory, which is operated as a scientific partnership among the California Institute of Technology, the University of California, and NASA; the observatory was made possible by the generous financial support of the W. M. Keck Foundation. KAIT has received donations from Sun Microsystems, Inc., the Hewlett-Packard Company, AutoScope Corporation, Lick Observatory, the National Science Foundation, the University of California, and the Sylvia and Jim Katzman Foundation. I am grateful to the editors of this volume for their incredible patience while waiting for my review.

References

Afshordi, N., Loh, Y.-S., & Strauss, M. S. 2004, *Phys. Rev. D*, **69**, 083524.
Aguirre, A. N. 1999a, *ApJ*, **512**, L19.
Aguirre, A. N. 1999b, *ApJ*, **525**, 583.
Aguirre, A. N., & Haiman, Z. 1999, *ApJ*, **525**, 583.
Aldering, G., Knop, R., & Nugent, P. 2000, *AJ*, **119**, 2110.
Bahcall, J. N., et al. 1996, *ApJ*, **457**, 19.
Bahcall, N. A., Ostriker, J. P., Perlmutter, S., & Steinhardt, P. J. 1999, *Science*, **284**, 1481.
Balbi, A., et al. 2000, *ApJ*, **545**, L1.
Barber, A. J., Thomas, P. A., Couchman, H. M. P., & Fluke, C. J. 2000, *MNRAS*, **319**, 267.
Barris, B., et al. 2004, *ApJ*, **602**, 571.
Benítez, N., Riess, A., Nugent, P., Dickinson, M., Chornock, R., & Filippenko, A. V. 2002, *ApJ*, **577**, L1.
Blakeslee, J. P., et al. 2003, *ApJ*, **589**, 693.
Boughn, S., & Crittenden, R. 2004, *Nature*, **427**, 45.
Branch, D. 1981, *ApJ*, **248**, 1076.
Branch, D. 1998, *ARA&A*, **36**, 17.
Branch, D., Baron, E., Thomas, R. C., Kasen, D., Li, W., & Filippenko, A. V. 2004b, *PASP*, in press (astro-ph/0408130).
Branch, D., Fisher, A., & Nugent, P. 1993, *AJ*, **106**, 2383.
Branch, D., & Miller, D. L. 1993, *ApJ*, **405**, L5.
Branch, D., Romanishin, W., & Baron, E. 1996, *ApJ*, **465**, 73 (erratum: **467**, 473).
Branch, D., & Tammann, G. A. 1992, *ARA&A*, **30**, 359.
Branch, D., et al. 2004a, *ApJ*, **606**, 413.
Caldwell, R. R., Davé, R., & Steinhardt, P. J. 1998, *Ap&SS*, **261**, 30.
Caldwell, R. R., Kamionkowski, M., & Weinberg, N. N. 2003, *Phys. Rev. Lett.*, **91**, 71301.
Cappellaro, E., Evans, R., & Turatto, M. 1999, *A&A*, **351**, 459.
Cappellaro, E., Turatto, M., Tsvetkov, D. Yu., Bartunov, O. S., Pollas, C., Evans, R., & Hamuy, M. 1997, *A&A*, **322**, 431.
Carroll, S. M., Hoffman, M., & Trodden, M. 2003, *Phys. Rev. D*, **68**, 023509.
Carroll, S. M., Press, W. H., & Turner, E. L. 1992, *ARA&A*, **30**, 499.
Chaboyer, B., Demarque, P., Kernan, P. J., & Krauss, L. M. 1998, *ApJ*, **494**, 96.
Coil, A. L., et al. 2000, *ApJ*, **544**, L111.
Cowan, J. J., McWilliam, A., Sneden, C., & Burris, D. L. 1997, *ApJ*, **480**, 246.
Dahlen, T., & Fransson, C. 1999, *A&A*, **350**, 349.
Dahlen, T., et al. 2004, *ApJ*, **613**, 189.
Davis, M., Newman, J. A., Faber, S. M., & Phillips, A. C. 2001, in *Deep Fields*, ed. S. Cristiani, A. Renzini, & R. E. Williams (Berlin: Springer), 241.
de Bernardis, P., et al. 2000, *Nature*, **404**, 955.
de Bernardis, P., et al. 2002, *ApJ*, **564**, 559.
Drell, P. S., Loredo, T. J., & Wasserman, I. 2000, *ApJ*, **530**, 593.
Efstathiou, G., et al. 1999, *MNRAS*, **303**, L47.
Efstathiou, G., et al. 2002, *MNRAS*, **330**, L29.
Eisenstein, D. J., Hu, W., & Tegmark, M. 1998, *ApJ*, **504**, L57.

Falco, E. E., et al. 1999, *ApJ*, **523**, 617.
Filippenko, A. V. 1997a, in *Thermonuclear Supernovae*, ed. P. Ruiz-Lapuente et al. (Dordrecht: Kluwer), 1.
Filippenko, A. V. 1997b, *ARA&A*, **35**, 309.
Filippenko, A. V. 2001, *PASP*, **113**, 1441.
Filippenko, A. V. 2003, in *From Twilight to Highlight: The Physics of Supernovae*, ed. W. Hillebrandt & B. Leibundgut (Berlin: Springer-Verlag), 171.
Filippenko, A. V., Li, W. D., Treffers, R. R., & Modjaz, M. 2001, in *Small-Telescope Astronomy on Global Scales*, ed. W. P. Chen, C. Lemme, & B. Paczyński (San Francisco: ASP), 121.
Filippenko, A. V., & Riess, A. G. 1998, *Phys. Rep.*, **307**, 31.
Filippenko, A. V., et al. 1992a, *AJ*, **104**, 1543.
Filippenko, A. V. 1992b, *ApJ*, **384**, L15.
Foley, R. J., et al. 2003, *PASP*, **115**, 1220.
Ford, C. H., et al. 1993, *AJ*, **106**, 1101.
Fosalba, P., et al. 2003, *ApJ*, **597**, L89.
Freedman, W., et al. 2001, *ApJ*, **553**, 47.
Gal-Yam, A., & Maoz, D. 2004, *MNRAS*, **347**, 942.
Garnavich, P., et al. 1998a, *ApJ*, **493**, L53.
Garnavich, P., et al. 1998b, *ApJ*, **509**, 74.
Garnavich, P., et al. 2002, *BAAS*, **34**, 1233.
Giavalisco, M., et al. 2002, *IAUC 7981*.
Gibson, B. K., et al. 2000, *ApJ*, **529**, 723.
Gilliland, R. L., & Phillips, M. M. 1998, *IAUC 6810*.
Goldhaber, G., & Perlmutter, S. 1998, *Phys. Rep.*, **307**, 325.
Goldhaber, G., et al. 1997, in *Thermonuclear Supernovae*, ed. P. Ruiz-Lapuente et al. (Dordrecht: Kluwer), 777.
Goldhaber, G., et al. 1998a, *BAAS*, **30**, 1325.
Goldhaber, G., et al. 1998b, in *Gravity: From the Hubble Length to the Planck Length*, SLAC Summer Institute (Stanford, CA: SLAC).
Goldhaber, G., et al. 2001, *ApJ*, **558**, 359.
Goobar, A., Bergstrom, L., & Mortsell, E. 2002, *A&A*, **384**, 1.
Goobar, A., & Perlmutter, S. 1995, *ApJ*, **450**, 14.
Gratton, R. G., Fusi Pecci, F., Carretta, E., Clementini, G., Corsi, C. E., & Lattanzi, M. 1997, *ApJ*, **491**, 749.
Groom, D. E. 1998, *BAAS*, **30**, 1419.
Hamuy, M., Phillips, M. M., Maza, J., Suntzeff, N. B., Schommer, R. A., & Aviles, R. 1995, *AJ*, **109**, 1.
Hamuy, M., Phillips, M. M., Maza, J., Suntzeff, N. B., Schommer, R. A., & Aviles, R. 1996a, *AJ*, **112**, 2391.
Hamuy, M., Phillips, M. M., Maza, J., Suntzeff, N. B., Schommer, R. A., & Aviles, R. 1996b, *AJ*, **112**, 2398.
Hamuy, M., Phillips, M. M., Maza, J., Suntzeff, N. B., Schommer, R. A., & Aviles, R. 1996c, *AJ*, **112**, 2408.
Hamuy, M., Trager, S. C., Pinto, P. A., Phillips, M. M., Schommer, R. A., Ivanov, V., & Suntzeff, N. B. 2000, *AJ*, **120**, 1479.
Hanany, S., et al. 2000, *ApJ*, **545**, L5.
Hancock, S., Rocha, G., Lazenby, A. N., & Gutiérrez, C. M. 1998, *MNRAS*, **294**, L1.
Hardin, D., et al. 2000, *A&A*, **362**, 419.
Hatano, K., Branch, D., & Deaton, J. 1998, *ApJ*, **502**, 177.
Höflich, P., Wheeler, J. C., & Thielemann, F. K. 1998, *ApJ*, **495**, 617.

Holz, D. E. 1998, *ApJ*, **506**, L1.
Holz, D. E., & Wald, R. 1998, *Phys. Rev. D*, **58**, 063501.
Hoyle, F., Burbidge, G., & Narlikar, J. V. 2000, *A Different Approach to Cosmology* (Cambridge: Cambridge Univ. Press).
Ivanov, V. D., Hamuy, M., & Pinto, P. A. 2000, *ApJ*, **542**, 588.
Jha, S. 2002, *Ph.D. thesis*, Harvard University.
Kaiser, N., et al. 2002, *BAAS*, **34**, 1304.
Kallosh, R., & Linde, A. 2003, *J. Cosmology Astropart. Phys.*, **2**, 2.
Kantowski, R. 1998, *ApJ*, **507**, 483.
Kantowski, R., Vaughan, T., & Branch, D. 1995, *ApJ*, **447**, 35.
Kim, A., Goobar, A., & Perlmutter, S. 1996, *PASP*, **108**, 190.
Knop, R., et al. 2003, *ApJ*, **598**, 102.
Krisciunas, K., Phillips, M. M., & Suntzeff, N. 2004, *ApJ*, **602**, L81.
Krisciunas, K., et al. 2001, *AJ*, **122**, 1616.
Krisciunas, K., et al. 2003, *AJ*, **125**, 166.
Krisciunas, K., et al. 2004, *AJ*, **127**, 1664.
Leibundgut, B. 2001, *ARA&A*, **39**, 67.
Leibundgut, B., et al. 1993, *AJ*, **105**, 301.
Leibundgut, B., et al. 1996, *ApJ*, **466**, L21.
Leonard, D. C., et al. 2002a, *PASP*, **114**, 35 (erratum: **114**, 1291).
Leonard, D. C., et al. 2002b, *AJ*, **124**, 2490.
Li, W., Filippenko, A. V., Chornock, R., & Jha, S. 2003a, *ApJ*, **586**, L9.
Li, W., Filippenko, A. V., Chornock, R., & Jha, S. 2003c, *PASP*, **115**, 844.
Li, W., Filippenko, A. V., Treffers, R. R., Riess, A. G., Hu, J., & Qiu, Y. 2001b, *ApJ*, **546**, 734.
Li, W., et al. 2000, in *Cosmic Explosions*, ed. S. S. Holt & W. W. Zhang (New York: AIP), 103.
Li, W., et al. 2001a, *PASP*, **113**, 1178.
Li, W., et al. 2003b, *PASP*, **115**, 453.
Lineweaver, C. H. 1998, *ApJ*, **505**, L69.
Lineweaver, C. H., & Barbosa, D. 1998, *ApJ*, **496**, 624.
Livio, M. 2001, in *Supernovae and Gamma-Ray Bursts: The Greatest Explosions since the Big Bang*, ed. K. Sahu, M. Livio, & N. Panagia (Cambridge: Cambridge Univ. Press), 334.
Madau, P., Della Valle, M., & Panagia, N. 1998, *MNRAS*, **297**, L17.
Maoz, D., & Gal-Yam, A. 2004, *MNRAS*, **347**, 951.
Matheson, T., Filippenko, A. V., Li, W., Leonard, D. C., & Shields, J. C. 2001, *AJ*, **121**, 1648.
Matheson, T., et al. 2003, *ApJ*, **599**, 394.
Matteucci, F., & Recchi, S. 2001, *ApJ*, **558**, 351.
Modjaz, M., Li, W., Filippenko, A. V., King, J. Y., Leonard, D. C., Matheson, T., Treffers, R. R., & Riess, A. G. 2001, *PASP*, **113**, 308.
Narlikar, J. V., & Arp, H. C. 1997, *ApJ*, **482**, L119.
Netterfield, C. B., et al. 2002, *ApJ*, **571**, 604.
Nolta, M. R., et al. 2004, *ApJ*, **608**, 10.
Nomoto, K., Umeda, H., Hachisu, I., Kato, M., Kobayashi, C., & Tsujimoto, T. 2000, in *Type Ia Supernovae: Theory and Cosmology*, ed. J. C. Niemeyer & J. W. Truran (Cambridge: Cambridge Univ. Press), 63.
Norgaard-Nielsen, H., et al. 1989, *Nature*, **339**, 523.
Nugent, P., 2001, in *Particle Physics and Cosmology: Second Tropical Workshop*, ed. J. F. Nieves (New York: AIP), 263.
Nugent, P., Kim, A., & Perlmutter, S. 2002, *PASP*, **114**, 803.
Nugent, P., Phillips, M., Baron, E., Branch, D., & Hauschildt, P. 1995, *ApJ*, **455**, L147.
Ostriker, J. P., & Steinhardt, P. J. 1995, *Nature*, **377**, 600.

Oswalt, T. D., Smith, J. A., Wood, M. A., & Hintzen, P. 1996, *Nature*, **382**, 692.
Paerels, F., Petric, A., Telis, G., & Helfand, D. J. 2002, *BAAS*, **34**, 1264.
Pain, R., et al. 1996, *ApJ*, **473**, 356.
Pain, R., et al. 2002a, *ApJ*, **577**, 120.
Pain, R., et al. 2002b, *BAAS*, **34**, 1169.
Parodi, B. R., et al. 2000, *ApJ*, **540**, 634.
Peacock, J. A., et al. 2001, *Nature*, **410**, 169.
Percival, W., et al. 2001, *MNRAS*, **327**, 1297.
Perlmutter, S., et al. 1995a, *ApJ*, **440**, L41.
Perlmutter, S., et al. 1995b, *IAUC 6270.*
Perlmutter, S., et al. 1997, *ApJ*, **483**, 565.
Perlmutter, S., et al. 1998, *Nature*, **391**, 51.
Perlmutter, S., et al. 1999, *ApJ*, **517**, 565.
Phillips, M. M. 1993, *ApJ*, **413**, L105.
Phillips, M. M., et al. 1992, *AJ*, **103**, 1632.
Phillips, M. M., et al. 1999, *AJ*, **118**, 1766.
Pskovskii, Yu. P. 1977, *Sov. Astron.*, **21**, 675.
Pskovskii, Yu. P. 1984, *Sov. Astron.*, **28**, 658.
Reiss, D. 2000, *PhD thesis*, University of Washington.
Riess, A. G., Filippenko, A. V., Li, W. D., & Schmidt, B. P. 1999b, *AJ*, **118**, 2668.
Riess, A. G., Nugent, P. E., Filippenko, A. V., Kirshner, R. P., & Perlmutter, S. 1998a, *ApJ*, **504**, 935.
Riess, A. G., Press, W. H., & Kirshner, R. P. 1995, *ApJ*, **438**, L17.
Riess, A. G., Press, W. H., & Kirshner, R. P. 1996a, *ApJ*, **473**, 88.
Riess, A. G., Press, W. H., & Kirshner, R. P. 1996b, *ApJ*, **473**, 588.
Riess, A. G., et al. 1997, *AJ*, **114**, 722.
Riess, A. G., et al. 1998b, *AJ*, **116**, 1009.
Riess, A. G., et al. 1999a, *AJ*, **117**, 707.
Riess, A. G., et al. 1999c, *AJ*, **118**, 2675.
Riess, A. G., et al. 2000, *ApJ*, **536**, 62.
Riess, A. G., et al. 2001, *ApJ*, **560**, 49.
Riess, A. G., et al. 2004a, *ApJ*, **600**, L163.
Riess, A. G., et al. 2004b, *ApJ*, **607**, 665.
Ruiz-Lapuente, P., et al. 1992, *ApJ*, **387**, L33.
Saha, A., et al. 1997, *ApJ*, **486**, 1.
Saha, A., et al. 2001, *ApJ*, **562**, 314.
Salpeter, E. E. 1955, *ApJ*, **121**, 161.
Sandage, A., & Tammann, G. A. 1993, *ApJ*, **415**, 1.
Sandage, A., et al. 1996, *ApJ*, **460**, L15.
Schmidt, B. P., et al. 1998, *ApJ*, **507**, 46.
Scranton, R., et al. 2004, *Phys. Rev. Lett.*, submitted (astro-ph/0307335).
Smith, R. C., et al. 2002, *BAAS*, **34**, 1232.
Spergel, D. N., et al. 2003, *ApJS*, **148**, 175.
Strolger, L. 2003, *PhD thesis*, University of Michigan.
Strolger, L., et al. 2004, *ApJ*, **613**, 200.
Sullivan, M., et al. 2003, *MNRAS*, **340**, 1057.
Suntzeff, N. 1996, in *Supernovae and Supernova Remnants*, ed. R. McCray & Z. Wang (Cambridge: Cambridge Univ. Press), 41.
Suntzeff, N., et al. 1996, *IAUC 6490.*
Tonry, J. L., et al. 2003, *ApJ*, **594**, 1.

Tripp, R. 1997, *A&A*, **325**, 871.
Tripp, R. 1998, *A&A*, **331**, 815.
Turatto, M., et al. 1996, *MNRAS*, **283**, 1.
Turner, M. S., & Riess, A. G. 2002, *ApJ*, **569**, 18.
Tyson, J. A., & Angel, R. 2001, in *The New Era of Wide Field Astronomy*, ed. R. Clowes, et al. (San Francisco: ASP), 347.
Umeda, H., et al. 1999, *ApJ*, **522**, L43.
van den Bergh, S., & Pazder, J. 1992, *ApJ*, **390**, 34.
Vaughan, T. E., Branch, D., Miller, D. L., & Perlmutter, S. 1995, *ApJ*, **439**, 558.
Wambsganss, J., Cen, R., & Ostriker, J. P. 1998, *ApJ*, **494**, 29.
Wang, Y., & Tegmark, M. 2004, *Phys. Rev. Lett.*, **92**, 241302.
Williams, B., et al. 2003, *AJ*, **126**, 2608.
Wilson, G., Cowie, L. L., Barger, A. J., & Burke, D. J. 2002, *AJ*, **124**, 1258.
Wright, E. L. 2001, *BAAS*, **34**, 574.
Yungelson, L. R., & Livio, M. 2000, *ApJ*, **528**, 108.
Zaldarriaga, M., Spergel, D. N., & Seljak, U. 1997, *ApJ*, **488**, 1.

THE PROGENITORS OF TYPE IA SUPERNOVAE

Review paper

Christopher A. Tout
University of Cambridge, Institute of Astronomy,
The Observatories, Madingley Road, Cambridge CB3 0DS, England
cat@ast.cam.ac.uk

Abstract Type Ia supernovae are identified as exploding degenerate stars. Their luminosity is due to the radioactive decay of about a solar mass of ^{56}Ni through ^{56}Co to ^{56}Fe. As such they are a major source of iron in the inter-stellar medium. Although it is generally accepted that a degenerate carbon/oxygen white dwarf explodes as it accretes material from a binary companion, the progenitors of type Ia supernovae have not been categorically identified. We discuss the various possible progenitors in detail and indicate theoretical and observational difficulties with each possibility. It may well be that the true nature of the progenitors has not yet even been conceived of. We look at why population synthesis fails to help distinguish and consider how the advent of population nucleosynthesis may change this. When used as universal standard candles SNe Ia are calibrated with the Phillips relation between absolute luminosity and light curve shape. This must therefore be valid at all redshifts and so both the absolute luminosity and the light curve decay must only depend on a single major property of the progenitors. We report on the latest understanding of this relation and find little to justify its universality beyond the local empirical evidence. A major effect on the absolute luminosities is the neutron to proton ratio at the time of the explosion because this determines the fraction of iron group elements made up of ^{56}Ni.

Keywords: Supernovae

1. Introduction

Luminous type Ia supernovae (hereinafter SNe Ia) are amongst the brightest objects in the Universe. Observations indicate that their absolute magnitudes lie in a narrow range, $M_{\rm B} = -18.5\pm0.5$, or $\pm60\%$ in luminosity (from Table 1 of Rowan-Robinson, 2002). Furthermore observations reveal a correlation between maximum absolute brightness and light curve shape (Phillips 1993) that facilitates an effective reduction in the standard deviation of absolute luminosities to $\pm15\%$. This small spread, coupled with the fact that they can be

E.M. Sion, S. Vennes and H.L. Shipman (eds.), White Dwarfs: Cosmological and Galactic Probes, 135-151.
© 2005 *Springer.*

seen to great distances make SNe Ia excellent standard candles for study of the cosmology of the Universe. Observations have, quite precisely, determined the rate of expansion, immortalised in the Hubble constant, (Branch 1998) and have further determined that this rate is accelerating with time, a measurement that has led to the invocation of a cosmological constant contributing about 70% of the critical density of the Universe to be added to matter's contribution of 30% (Perlmutter et al. 1999, Riess et al. 1998). Further credence is lent to this result by cold-dark-matter models of the angular variation of the cosmic microwave background (de Bernardis et al. 2000) and gravitational lensing measurements (Wittman et al. 2000). However the supernovae result could easily be misleading. If SNe Ia at a redshift of $z \approx 1$ happen to be as little as 25% fainter than locally a cosmology with no cosmological constant can be fitted. Such a small effect would be lost in the uncorrected spread of 60% so we rely on the empirical calibration between peak brightness and light curve shape.

Most agree that before we can be absolutely certain of the cosmological results we must identify the actual progenitors of the SNe. We review candidates put forward to date and reach the unfortunate conclusion that, in its simplest form, each has a good reason to be rejected. We add a few more suggestions but deduce that we most likely have not yet knowingly stumbled upon the actual progenitors. We describe how population synthesis, once the hope of many, is near useless on its own. Even when coupled with nucleosynthesis it is not as discriminating as had been hoped (Tout et al. 2001). We then consider the current status of the mechanism that underlies the Phillips relation and sadly deduce that this too is not understood.

2. Types of Supernovae

Supernovae are classified by their spectra. Those that show hydrogen lines are type II and those that do not are type I. The SNe I are further subdivided into the SNe Ia that show prominent silicon lines, the SNe Ib that show no Si lines and the SNe Ic that show no Si nor helium lines. The various types of SNe II are thought to be the deaths of massive stars whose degenerate cores have exceeded the Chandrasekhar mass and collapse to neutron stars. The gravitational energy released in such an event amounts to $E_{\rm B} \approx 3 \times 10^{46}$ J which would be $10^{13}\,L_\odot$ for three months. However most of this energy is released in neutrinos, only a few of which interact with the stellar material and eject the hydrogen-rich envelope. The SNe Ib/c are all associated with recent star formation and so are most probably the deaths of very massive stars that have lost their hydrogen envelopes in a phase of prolonged rapid mass loss before they explode. The SNe Ia are quite different. They are explained by the thermonuclear explosion of about a solar mass of degenerate material that is converted to ^{56}Ni and then expelled to the ISM. Their light curves show the decay of this via ^{56}Co to ^{56}Fe. There must be enough nuclear energy available to overcome the binding energy of the white dwarf. In practice about $0.8\,M_\odot$ of iron-group material, mostly in nuclear statistical equilibrium is expelled to the ISM. If $1\,M_\odot$ of material, originally one fifth carbon and four-fifths oxygen, is converted to ^{56}Fe then 1.8×10^{44} J of energy are available and the decay of this from ^{56}Ni releases 2×10^{43} J which is enough to power the supernova at $5 \times 10^9\,L_\odot$ for 80 d.

3. White-Dwarf Progenitors

White dwarfs may be divided into three major types: (i) helium white dwarfs, composed almost entirely of He, form as the degenerate cores of low-mass red giants ($M \leq 2\,M_\odot$) which lose their hydrogen envelope before helium can ignite; (ii) carbon/oxygen white dwarfs, composed of about 20% C and 80% O, form as the cores of asymptotic giant branch stars or naked helium burning stars that lose their envelopes before carbon ignition (with progenitors of typically $1 - 6\,M_\odot$); and (iii) oxygen/neon white dwarfs, composed of heavier combinations of elements, form from giants that ignite carbon in their cores but still lose their envelopes before the degenerate centre collapses to a neutron star (with progenitors of typically $6 - 8\,M_\odot$).

In a close binary system, mass transfer can increase the mass of a white dwarf. As its mass approaches the Chandrasekhar limit ($M_{\rm Ch} \approx 1.44\,M_\odot$) degeneracy pressure can no longer support the star which collapses releasing its gravitational energy. In ONe white dwarfs the collapse is hastened by electron captures on to magnesium and they lose enough energy in neutrinos to collapse sufficiently, before oxygen ignites, to avoid explosion (accretion induced collapse, AIC). The CO white dwarfs, on the other hand, reach temperatures early enough during collapse, typically at $1.38\,M_\odot$ if initially cold enough, for carbon fusion to set off a thermonuclear runaway under degenerate conditions and release enough energy to create a SN Ia. Accreting He white dwarfs reach sufficiently high temperatures to ignite helium well below $M_{\rm Ch}$ ($M \approx 0.7\,M_\odot$, Woosley, Taam & Weaver 1986) but an explosion under these conditions is expected to be quite unlike a SN Ia and the apparent lack of such objects suggests ignition is always gentle enough to avoid explosion.

4. The Standard Model

Accreting white dwarfs have been known for some time as the engines of cataclysmic variables, the source of novae and dwarf novae (Warner 1995) and so are the first candidate to be considered. However if the accreting material is hydrogen-rich, accumulation of a layer of only $10^{-4}\,M_\odot$ or so leads to ignition of hydrogen burning sufficiently violent to eject most, if not all of or more than, the accreted layer in the well known novae outbursts of cataclysmic variables. The white dwarf mass does not significantly increase and ignition of its interior is avoided. However if the accretion rate is high $\dot{M} > 10^{-7}\,M_{\odot}\,\mathrm{yr}^{-1}$ hydrogen can burn as it is accreted, bypassing novae explosions (Paczyński & Żytkow 1978), and allowing the white dwarf mass to grow. Though, if it is not much larger than this, $\dot{M} > 3 \times 10^{-7}\,M_{\odot}\,\mathrm{yr}^{-1}$, hydrogen cannot burn fast enough so that accreted material builds up a giant-like envelope around the core and burning shell which eventually leads to more drastic interaction with the companion and the end of the mass transfer episode. Rates in the narrow range for steady burning are found only when the companion is in the short-lived phase of thermal-timescale expansion as it evolves from the end of the main sequence to the base of the giant branch. Super-soft X-ray sources (Kahabka & van den Heuval 1997) are probably in such a state but cannot be expected to remain in it for very long and white dwarf masses almost never increase sufficiently to explode as SNe Ia.

The standard model overcomes this problem by postulating that, when the mass-transfer rate exceeds that allowed for steady burning, only just the right fraction of the mass transferred is actually accreted by the white dwarf. The standard mechanism currently invoked is a strong wind from the accretion disc that expels the material from the system before it reaches the white dwarf (Hachisu, Kato & Nomoto 1996). An alternative might be that the white dwarf does indeed swell up to giant dimensions but that the resulting common-envelope evolution is very efficient so that the small amount of excess material can be ejected without the cores spiralling in. This is quite consistent with the findings of Nelemans & Tout (2004; Nelemans et al. 2000) who find such efficiency necessary for at least one phase of common-envelope evolution in the formation of close double white dwarf systems.

5. Merging CO White Dwarfs

A more promising scenario is mass transfer from one white dwarf to another. In a very close binary orbit gravitational radiation can drive two white dwarfs together until the less massive fills its Roche lobe. If both white dwarfs are CO and their combined mass exceeds M_{Ch} enough mass could be transferred to set off a SN Ia. However if the mass ratio $M_{\mathrm{donor}}/M_{\mathrm{accretor}}$ exceeds 0.628 mass transfer is dynamically unstable because a white dwarf expands as it loses mass. Based on calculations at somewhat lower, steady accretion rates, Nomoto and Iben (1985) have shown that the ensuing rapid accretion of material allows carbon to burn in mild shell flashes converting the white dwarf to ONe and ultimately leading to AIC and not a SN Ia. We have ourselves confirmed this using the Eggleton (1971) evolution code. For smaller mass ratios accretion proceeds on the gravitational radiation timescale but even this is fast

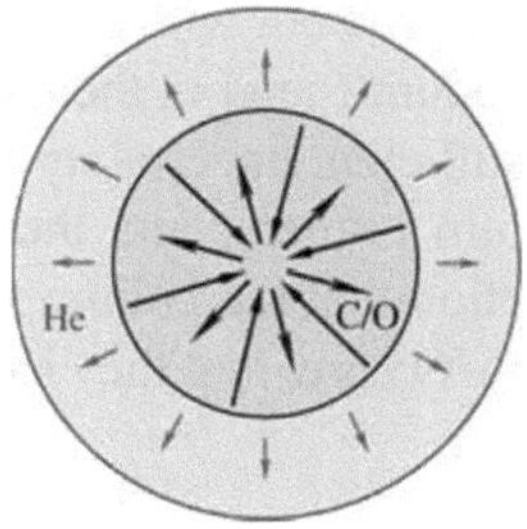

Figure 1. A schematic diagram of and edge-lit detonation (ELD). A degenerate CO white dwarf accretes a degenerate layer of He-rich material. When the base of this layer reaches a sufficiently high temperature the triple-α reaction ignites explosively. The shock of this explosion sets off explosive carbon burning in the CO core. Most of both the CO mixture and the helium shell are burnt to ^{56}Ni, while the high fraction of He nuclei in the outer layer leads to an α-rich freezeout creating such nuclei as ^{44}Ti.

enough to allow gentle burning of the carbon. This model too can be saved by postulating a mechanism to limit the accretion rate. A popular possibility is rapid rotation of the stellar surface (Yoon & Langer 2004).

6. Edge-Lit Detonations

If a CO white dwarf accretes from a He white dwarf (or a naked helium burning star) the mass ratio is generally small enough for dynamically stable mass transfer so that a helium layer builds up on the surface of the CO white dwarf. As in the nova explosions of hydrogen, the base of this helium layer eventually reaches a temperature at which helium can ignite in a degenerate flash. Unlike the novae this requires about $0.15\,M_\odot$ of helium (Woosley & Weaver 1994). Modelling this process with the Eggleton evolution code, we find that the actual mass of helium required depends both on the mass of the CO core and the rate of accretion. It ranges from about $0.2\,M_\odot$ for low-mass cores accreting slowly to as little as $0.02\,M_\odot$ for high-mass cores accreting rapidly. It can easily be envisaged that ignition of such a helium layer can detonate the CO core either by compressing it or by an inwardly propagating heating front (Fig. 1). Typical total masses of these edge-lit detonations (hereinafter ELDs) are again of the order of a solar mass so enough energy exists to explode the star and enough ^{56}Ni can be synthesized to power a SN Ia. However Kawai, Saio & Nomoto (1987) investigated helium accreting spherically on to a CO core and claimed that helium can burn non-degenerately if accreted steadily at rates above about $3 \times 10^{-8}\,M_\odot\,\mathrm{yr}^{-1}$ and so avoid setting off an ELD. Our own calculations support this conclusion and because gravitational radiation drives mass transfer from a He white dwarf donor at a rate considerably in excess of this, progenitors involving two white dwarfs are ruled out. However the mass transfer rate from naked helium stars is small enough to avoid such non-degenerate burning. A further problem with any claim that the majority of SNe Ia are ELDs is the lack of helium found in their spectra (Mazzali & Lucy 1998). However there remains sufficient uncertainty in the explosion models, such as exactly how much helium is there, and the conversion to observed spectra, in particular the assumption that the helium shell remains spherically symmetric around the exploding CO core, that they cannot be ruled out on this ground alone.

7. Oxygen Neon White Dwarfs with Unburnt Cores

Super AGB stars are defined to be those that undergo carbon ignition before losing their hydrogen envelopes to leave an ONe white dwarf. In the lowest-mass SAGB stars this ignition begins degenerately and off centre. Some models, including our own when the resolution is low, appeared to leave a small unburnt CO core at the centre of a cool ONe white dwarf. The carbon in such a core could ignite and set off a thermonuclear runaway through the whole white dwarf as it accretes material and approaches the Chandrasekhar mass and so avoid AIC. This would have the added benefit that these ONe white dwarfs would be of higher mass than their CO counterparts and so would require much less mass transfer. This is however almost certainly a numerical artifact. Models with well-resolved cores always burn carbon right to their centres (Siess, private communication).

8. Long-Period Dwarf Novae

Another attempt to save cataclysmic variables as progenitors has been made recently by King, Rolfe & Schenker (2003). They noted that long-period dwarf novae, in which the companion to the white dwarf is a low-mass red giant, transfer mass at an average rate of $10^{-8}\,M_{\odot}\,\mathrm{yr}^{-1}$. However, because in dwarf novae most of the mass is transferred in outbursts that last only about one tenth of the inter-outburst time, the actual accretion rate is raised to $10^{-7}\,M_{\odot}\,\mathrm{yr}^{-1}$, sufficient to avoid nova explosions. They would spend long enough in this high state to accrete about $0.4\,M_{\odot}$ in total. The white dwarf would need to start off with a mass of $1\,M_{\odot}$ or more and this would have been achieved in a previous thermal-timescale mass-transfer phase as a super-soft source. Though the idea is promising it fails because the critical factor affecting novae is the cooling of the accreted material so that it becomes degenerate. This cooling can of course take place during the long quiescent phases unaffected by the brief intervals of rapid accretion.

9. Single Star Progenitors

Single star models of SNe Ia are generally ruled out on two counts. First because no star is expected to reach the required state according to standard stellar evolution and secondly because, if any single star naturally reached such a state, there ought to be far more SNe Ia than are seen. Convoluted binary pathways lead to a natural scarcity of any particular outcome.

However if a $1.5\,M_{\odot}$ star were to evolve without mass loss at all its core would eventually approach the Chandrasekhar mass in a degenerate state and a SN Ia would result. But there is no reasonable mechanism to prevent mass loss, particularly on the AGB when the envelope is only loosely bound and thermal pulses slow down core growth. There does though remain the remote possibility that a rare stellar property, perhaps an intense primordial magnetic field, might inhibit mass loss totally.

Another possibility (Fig. 2) is a $7\,M_{\odot}$ star that would normally pass through an SAGB phase but that loses its envelope just before second dredge up. The star in the Figure has been evolved without mass loss to the AGB when the deep convective envelope reaches down to the hydrogen-burning shell for the second time. If a typical low rate of mass loss were to continue the helium-burning shell would increase in luminosity to such an extent that the hydrogen-burning shell would expand, cool and extinguish. The convective envelope would then carry fresh hydrogen all the way down to the helium shell at less than $1\,M_{\odot}$. From then thermal pulses would begin and the core would not grow much more before the envelope were lost. However in this case, just at the start of the AGB, we have removed all the hydrogen envelope in a sudden burst of mass loss. Second dredge up is avoided and the CO core can then grow until carbon ignites in the centre degenerately.

All of this evolution is highly unusual. Even with a binary companion it is hard to see how the envelope can be removed just at the right moment. After all the star was larger when it first ascended the red giant branch and would somehow need to avoid mass transfer then. It would also need to lose no further mass as a naked helium star which is contrary to our understanding of Wolf-Rayet stars. Yet there remain many pathways that we have not yet investigated and the true progenitors may yet surprise us!

10. The Case of 2002ic

Recently many of our ideas were turned on their heads by observations of SN 2002ic (Hamuy et al. 2003). This was an apparently normal SN Ia showing clear silicon lines and the decay of $0.5 - 0.8\,M_{\odot}$ of ^{56}Ni. But it also showed hydrogen in its spectrum, and not just a little but $2 - 3\,M_{\odot}$! Livio & Riess (2003) suggested that it was a case of merging CO white dwarfs in a common envelope but all the problems with rapid accretion leading to non-degenerate ignition would only be exacerbated. Was it just a rare case of the standard model in which the circumstellar hydrogen is still around, perhaps? Or could

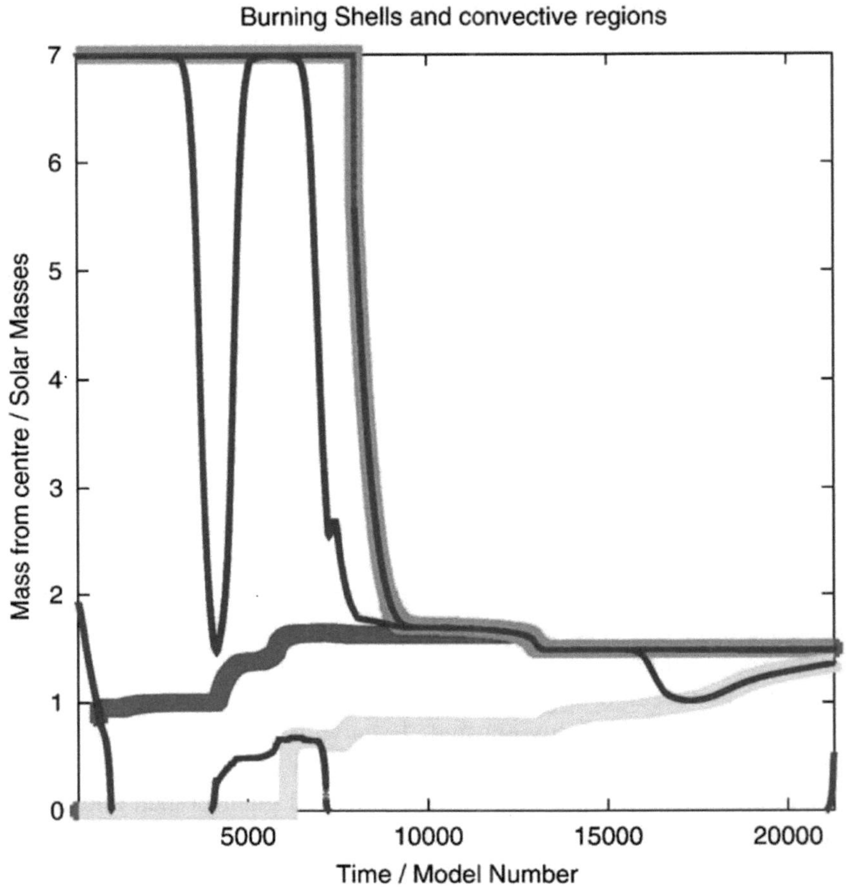

Figure 2. The internal structure of a 7 $M_\odot$ star plotted as mass against model number. Model number increases monotonically with time but there are more models when evolution is slower. Thus the longest-lived, main-sequence phase is confined to less than the first 1,000 models. The thin black lines are convective boundaries. The core convective region on the main sequence is visible at the left. This is replaced by a convective envelope on the first giant branch and then core convection during helium burning etc. The palest thick line indicates the point at which helium abundance has fallen below one tenth and so follows the helium burning shell. The darkest thick line similarly follows the hydrogen burning shell while the intermediate thick line is the total mass of the star. In this case a drastic phase of mass loss takes place just after the star joins the AGB, before second dredge up. The core can then grow until carbon ignites degenerately at the far right.

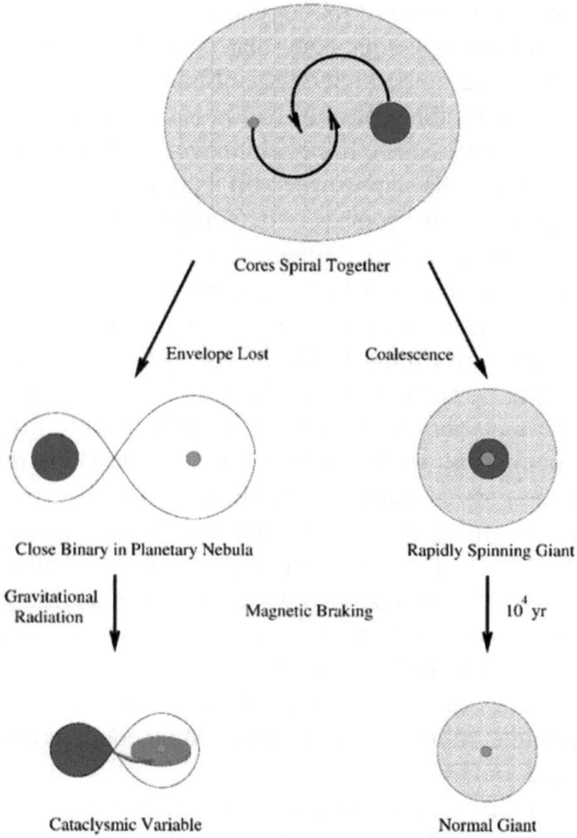

Figure 3. Common-envelope evolution. After dynamical mass transfer from a giant, a common envelope enshrouds the relatively dense companion and the core of the original giant. These two spiral together as their orbital energy is transferred to the envelope until either the entire envelope is lost or they coalesce. In the former case a close white-dwarf and main-sequence binary is left, initially as the core of a planetary nebula. Magnetic braking or gravitational radiation may shrink the orbit and create a cataclysmic variable. Coalescence results in a rapidly rotating giant which will very quickly spin down by magnetic braking.

it be a $4\,M_\odot$ star that evolved without mass loss? For now it remains an elusive piece of the jigsaw that cannot be discarded.

11. Population Synthesis

Given an initial state for a binary system, its masses M_1 and M_2 and its period[1] P, and a model for all the physical processes of stellar evolution and binary interaction, we can determine whether or not a particular system would end its life as a SN Ia. These physical processes include mass transfer both by Roche-lobe overflow or in a stellar wind, common-envelope evolution, magnetic and gravitational-radiation braking and all the associated effects on the evolution of the individual components that comprise the system. Unfortunately many of them are not sufficiently well understood for a precise quantification. Particularly troublesome for SNe Ia is common-envelope evolution (Fig. 3) because it is necessary to bring white dwarfs close enough to interact.

When a red giant or AGB star grows to fill its last stable potential surface or Roche lobe it begins to transfer mass to its companion. But, as it loses mass, the convective envelope of a giant expands. If it is still the more massive component of the binary, and mass transfer is conservative, the orbit and Roche lobe shrink. Consequently the process of mass transfer leads, on a dynamical timescale, to the giant overfilling its Roche lobe yet more. The overflow rate rapidly rises and the companion, typically a lower-mass main-sequence star, cannot accrete the material. Its own Roche lobe is quickly filled and a common envelope engulfs the whole system. The two cores, the relatively dense companion and the core of the giant are then assumed to spiral together by some, as yet undetermined, frictional mechanism. Some fraction of the orbital energy released is available to drive off the envelope. If all of it is ejected while the cores are still well separated we are left with a closer binary system comprising the unscathed companion and a white dwarf which may evolve to a cataclysmic variable. Alternatively, if some of the envelope still remains when the companion reaches the denser depths of the common envelope, they can merge leaving a single, rapidly rotating, giant. Magnetic braking quickly spins down these merged giants.

Our ignorance of this process is encapsulated in a constant $\alpha_{\rm CE}$ which measures the fraction of the released orbital energy that goes into driving off the envelope. Very few agree on the precise definition of $\alpha_{\rm CE}$ and its numerical value is uncertain to within a factor of ten. Because additional energy is always available from the thermal reservoir provided by the nuclear burning luminosity it may well exceed unity. Furthermore it is almost certainly not constant from one case to another.

Any white dwarf accreting matter from its companion must have passed through at least one phase of common-envelope evolution because the white dwarf must itself have been the core of a giant that could only have evolved in a much wider system. Many double degenerate systems have passed through two phases. To examine how uncertain are the predictions we divide the exploding white dwarfs into three types, (i) exploding Chandrasekhar-mass CO white dwarfs, (ii) ELDs and (iii) exploding He white dwarfs. Each type separates into groups of low and high initial-period systems. The high-period systems experience two common-envelope phases. The first, when the primary fills its Roche lobe as a giant, leaves the binary wide enough for the secondary also to evolve to a giant before it too fills its Roche lobe. The low-period systems experience only the second phase. Mass transfer from the primary begins while it is crossing the Hertzsprung gap, or perhaps while it is still on the main sequence. Mass transfer then proceeds only on a thermal timescale. After this Algol phase of evolution we are left with a white dwarf companion to a more massive secondary in a much wider system. The secondary then evolves to fill its Roche lobe only as a giant. Most of the progenitors fall into one of these two categories but some follow a considerably more convoluted evolution. We use the binary evolution model of Hurley, Tout and Pols (2002) who describe all the details of the evolutionary processes included in the model.

A lack of understanding of common-envelope evolution alone ought to be enough to make us very wary of the results of population synthesis. But, like others before us, we go further and calculate the SNe Ia rate for the various progenitors in a typical galaxy like our own. To do this we convolve a grid of models (about $5{,}000{,}000$ are needed to sufficiently resolve the $M_1 - M_2 - P$ space) with initial mass functions for each of the components, an initial period distribution, a binary fraction, a star formation history and a galactic model. Although each of these might be reasonably guessed on its own, together they give us enough freedom to achieve almost any result we want even when we require the model to fit observational constraints on all types of binary star, Algols, cataclysmic variables, X-ray binaries, symbiotics *etc.* Thus we stress that the results of binary population synthesis should not be glibly trusted.

12. The Supernova Rate

The observed rate of SNe Ia is 4 ± 2 per $1{,}000\,\mathrm{yr}$ per galaxy like our own (Cappellaro et al. 1997). Table 1 gives the rate for various possible progenitors for two of our population syntheses that differ only in the value of α_{CE}. We have chosen the primary mass M_1 from the mass function of Kroupa, Tout and Gilmore (1993), the secondary mass M_2 so as to give a uniform distribution of mass ratio $q = M_2/M_1$ and the semi-latera recta l from a flat distribution in $\log l$ with $3 < l/R_\odot < 3 \times 10^6$. The binary fraction and galactic model are condensed into the statement that one binary system with $M_1 \geq 0.8\,M_\odot$ forms per year per galaxy, as is appropriate for our own (Hurley, Tout & Pols 2002). The three major groups of progenitors can be further subdivided each into two distinct channels. The exploding Chandrasekhar-mass CO white dwarfs can be split into those that have dynamically unstable (coalescence) and those that have dynamically stable (accretion) mass transfer. The ELDs can be split into those CO white dwarfs accreting from a He white dwarf and those accreting from a naked helium-burning star. The exploding He white dwarfs, which we recall are not likely SNe Ia, can again be divided into those accreting from a white dwarf and those accreting from a naked helium star.

It is immediately apparent that by varying α_{CE} alone we can fit the observed rate with whichever subset of the progenitors we please. Because of

Table 1. Supernovae rates for various progenitors per 1,000 yr and their yields.

	$\alpha_{CE} = 1$	$\alpha_{CE} = 3$	$M_{Ni}/M_{\odot}$	$M_{Ti}/M_{\odot}$	M_{Ca}/M_{Fe}
Coalescence CO + CO	0.05	1.20	0.76	2.2×10^{-6}	1.9×10^{-4}
CO on to CO accretion	0.04	0.36	0.23	6.6×10^{-7}	1.9×10^{-4}
CO + He wd ELD	2.90	8.79	5.07	3.6×10^{-2}	5.1×10^{-3}
CO + naked He ELD	0.36	1.12	0.71	4.3×10^{-3}	4.6×10^{-3}
He wd + He wd ignition	0.15	1.39	0.63	1.2×10^{-2}	(1.3×10^{-2})
He wd + naked He ignition	0.00	0.28	0.13	2.5×10^{-3}	(1.3×10^{-2})
Total CO > M_{Ch}	0.09	1.57	0.99	2.8×10^{-6}	1.9×10^{-4}
Total ELD	3.26	9.91	5.79	4.0×10^{-2}	5.0×10^{-3}
Total He wd ignition	0.15	1.67	0.75	1.5×10^{-2}	(1.3×10^{-2})
Total	3.50	13.15	7.52	5.5×10^{-2}	5.3×10^{-3}

the two common-envelope phases involved the only other systems to be so strongly affected by α_{CE} are the non-interacting double degenerate systems but their numbers are not yet sufficient to constrain α_{CE} independently (Maxted & Marsh 1999). The well-studied cataclysmic variables usually only experience a single common-envelope phase in their evolution and so their numbers, which are uncertain anyway, depend rather more weakly on α_{CE}. We reiterate that, even without varying the initial mass and period distributions, which would affect other types of binary, we are able to fit the SNe Ia with whichever subset of the progenitors we please. The problem is that there are too few observables to constrain the free parameters of the model.

13. Nucleosynthesis

If we turn to nucleosynthesis we can tap into many more observable quantities from stellar abundances of elements to meteoritic and terrestrial isotopic ratios. Unfortunately this is not without the expense of introducing new parameters to our model but we can reduce the ratio of parameters to observables. Of particular interest to us for supernovae are the iron group elements and heavier isotopes that are not synthesized in earlier phases of stellar evolution such as s-processing in AGB stars. The greatest uncertainty is introduced by our poor understanding of the mechanism by which SNe II explode. It is not certain how much of their iron core is trapped in or falls back on to the newly formed neutron star and how much is expelled to the ISM. Examination of the models of Woosley and Weaver (1995) reveals that an average of about $0.08\,M_{\odot}$ of Fe (including the ^{56}Ni) expelled to the ISM per SN II is a reasonable but rather uncertain estimate. Because ^{56}Ni decay is the major source of energy in the SNe Ia their iron production is much better determined. Exploding Chandrasekhar-mass CO white dwarfs give about $0.63\,M_\odot$ each (Thielemann, Nomoto & Yokoi 1986). The ELD production depends on both CO core mass and He envelope mass (at least one extra parameter for the model) but the results of Livne and Arnett (1995) can be fitted quite well by

$$M_{\rm Ni} = 0.75 - 3.0(M_{\rm CO} - 1.0\,M_\odot)^2/M_\odot, \tag{1}$$

where $M_{\rm Ni}$ is the total mass of ^{56}Ni, that will decay to ^{56}Fe produced by an ELD with CO core mass $M_{\rm CO}$. The mass of iron liberated by exploding He white dwarfs would again depend very much on their mass and accretion rate. An appropriate model (Woosley, Taam & Weaver 1986) gives $0.45\,M_\odot$ per explosion. We have included these yields in our population synthesis calculations and table 1 lists the yields per 1,000 yr per galaxy for the model with $\alpha_{\rm CE} = 3$. If an average SN II contributes about $0.08\,M_\odot$ and their rate is about thrice that of SNe Ia (Cappellaro et al. 1997) then the iron contribution of SNe Ia is about two and a half times that of SNe II whatever the true progenitor.

14. Explosive Helium Burning

What might distinguish ELDs from exploding Chandrasekhar-mass CO white dwarfs? In the latter the thermonuclear runaway is confined to the CO rich mixture where it produces the ^{56}Ni and the Si peculiar to SNe Ia spectra. In the ELDs it occurs both in the CO core and in the He-rich envelope. In this envelope a small but very significant fraction of nuclei do not reach the end of the α-burning chain but freezeout as the envelope expands. Of particular interest are the heavier isotopes such as ^{44}Ti and ^{48}Cr which are not readily produced elsewhere. We concentrate on ^{44}Ti that decays, via ^{44}Sc, to ^{44}Ca in the ISM. The mass ratio of ^{44}Ca/^{56}Fe in the Solar System is 1.2×10^{-3}. An average SN II yields at most a ratio of about 6×10^{-4} (Timmes et al. 1996) and an exploding Chandrasekhar-mass white dwarf only 3×10^{-5} (Livne & Arnett 1995). As Timmes et al. point out these two alone cannot account for the Solar-System abundance and we must turn to explosive He burning.

The model for the exploding He white dwarf that gave $0.45\,M_\odot$ of iron (Woosley, Taam and Weaver 1986) yields $8.9 \times 10^{-3}\,M_\odot$ of ^{44}Ti so that only one such explosion per eighteen SNe II or six SNe Ia would be enough to account for the Solar-System ^{44}Ca. On the other hand our calculations with the Eggleton code show that it is very easy to raise the degeneracy of a He white dwarf by accretion before the triple-α reaction begins. The white dwarf becomes a naked helium burning star and subsequently a CO white dwarf without any explosive helium burning. In addition there are no obvious observed candidates for such explosions, which ought to be almost as bright as SNe Ia and we might reasonably discount them altogether.

The contribution from ELDs, not surprisingly, depends, like the Fe yield, on both the CO core mass and the He envelope mass. A fit to the models of Livne and Arnett (1995) is

$$M_{\rm Ti} = \begin{cases} 0.0033 + 0.143(M_{\rm CO} - 0.8\,M_\odot)^2/M_\odot & M_{\rm CO} < 0.8\,M_\odot \\ 0.0033\,M_\odot & M_{\rm CO} \geq 0.8\,M_\odot, \end{cases} \tag{2}$$

where $M_{\rm Ti}$ is the total mass of ^{44}Ti, that will decay to ^{44}Ca produced by an ELD with CO core mass $M_{\rm CO}$. Again for the model with $\alpha_{\rm CE} = 3$, the mass of ^{44}Ti returned to the ISM by each of the various progenitor types is recorded in table 1 and the final column gives the Solar-System ratio if only that progenitor type is combined with thrice as many SNe II. ELDs give a ratio that is four times as large and so are unlikely to dominate while exploding Chandrasekhar-mass CO white dwarfs give a ratio over six times too small. If we exclude the CO white dwarfs accreting from He white dwarfs on the grounds that they can burn helium non-degenerately then the combination of the ELDs accreting from naked helium stars, all the exploding Chandrasekhar-mass CO white dwarfs and the SNe II, with no exploding He white dwarfs, give a ratio of 2.0×10^{-3} within a factor of two of the measured value. This would therefore be our favoured model based on these calculations. The progenitors of 40% of its SNe Ia are ELDs which are unlikely to be standard candles (Regős et al. 2000).

However this result depends critically on the ^{44}Ti produced by SNe II and unfortunately models of core-collapse supernovae are in a much less reliable state than those of SNe Ia. Though the procedure is promising it is at present as useless as standard population synthesis for determining the progenitors of SNe Ia. We have embarked on a full study of the effects of binary stars on nucleosynthesis (Izzard & Tout 2003). The goal is to increase the number of observations with which we can compare at a faster rate than the number of uncertainties which we must include in the model. Then a fully consistent model will become fruitful.

15. The Peak Luminosity

In order to predict its variation we must understand why the peak luminosity of SNe Ia varies. For some time this was put down to the variation of the C/O ratio in the progenitor white dwarfs, a result of the range in mass of their progenitors. However current explosion models (Röpke & Hillebrandt 2004) show that this makes little difference. Out to $0.8\,M_\odot$ the whole core burns to nuclear statistical equilibrium (NSE) in a deflagration which ends when the density drops to a point where a detonation begins. In degenerate matter this density does not depend on composition. There may be differences in the inner $0.2\,M_\odot$ where weak interactions are important but these are small and once the detonation begins there is only incomplete burning to silicon. NSE determines which of the iron-group isotopes are formed. If the number of protons present is comparable to the number of neutrons then ^{56}Ni dominates. As the relative number of neutrons N increases relative to the number of protons Z so the equilibrium moves to favour the more neutron-rich ^{54}Fe. Because ^{54}Fe is stable it cannot power the SN in the same way as ^{56}Ni and so as N rises relative to Z the SNe become fainter.

In general neutron-rich material is dominated by ^{23}Na from carbon burning and ^{22}Ne from CNO elements that have been processed during helium burning. During hydrogen burning 98% of CNO elements are converted to ^{14}N which then acquires two alpha particles during helium burning. Timmes, Brown & Truran (2003) show that the variation in metallicity in the local SNe Ia host galaxies ($1/3 < Z/Z_\odot < 3$) is just enough to account for the variations in the peak luminosities of the SNe. Low metallicity gives rise to fewer neutrons and so brighter SNe.

16. The Status of the Phillips Relation

So where does this leave the Phillips relation. Brighter SNe have broader, slower light curves. So the light curve shape must also depend on the mass of ^{56}Ni produced or at least on something upon which it directly depends. Arnett (1982) demonstrated that the light curve decay time

$$\tau_{\rm lc} \propto \kappa_{\rm opt}^{1/2} M_{\rm ej}^{3/4} E_{\rm k}^{-1/4}, \qquad (3)$$

where $\kappa_{\rm opt}$ is the opacity of the ejecta, $M_{\rm ej}$ its mass and $E_{\rm k}$ its kinetic energy.

All iron group elements have complex line structures that make the opacity difficult to estimate but it is unlikely to vary much. Mazzali et al. (2001) suggested that only $M_{\rm ej}$ varies because this would lead to the required correlation. However we have discussed how models find it to be constant and yet allow the peak luminosity to vary. This leaves only the kinetic energy. The more carbon relative to oxygen in the core the more total energy is available and so $E_{\rm k}$ increases with C/O (Röpke & Hillebrandt 2004). Now we find in our own stellar models that C/O ratios are generally larger in low metallicity progenitors. If we combine this with the result of the previous section, where we found the peak luminosity to increase at low metallicity, we have an anticorrelation with $\tau_{\rm lc}$ and we deduce that we do not yet see the full picture!

17. Conclusions

While the nature of the progenitors of SNe Ia remains elusive their use as cosmological standard candles can be called into question. We have examined various progenitor models, some standard and some not so standard, and found all to be lacking in something either theoretically or observationally or both.

We deduce that we may not yet have even conceived of the true nature of the progenitors of SNe Ia.

Binary population synthesis alone cannot distinguish the progenitors by comparing the predicted SNe Ia rate with the observed. Binary population synthesis incorporating nucleosynthesis shows some promise.

The explanation of the Phillips relation remains elusive too. If our discussion has any truth and metallicity falls with redshift we would expect distant SNe to be brighter but at the same time to have faster light curves. The Phillips relation would imply that they are even fainter and this would put them even further away so that the Universal acceleration would be even greater! There is still much to be done. In particular the importance of neutrons at the time of explosion makes it all the more important to understand the processes that lead to their production and destruction such as the convective URCA process (Lesaffre, Podsiadlowski & Tout 2005) that can convert newly formed ^{23}Na to the even more neutron rich ^{23}Ne at the onset of carbon ignition.

Acknowledgments

The author thanks Churchill College for a Fellowship. Many thanks also go to John Eldridge for investigating some strange single-star evolution scenarios, Lionel Siess for discussions on the super AGB stars and Paulo Mazzali for various conversations over time. This work forms part of the European Research Training Network on the Physics of Type Ia Supernova Explosions (http://www.mpa-garching.mpg.de/ rtn/).

Notes

1. In practice a binary star may have an initial eccentricity but, in general, tides circularise the orbit before significant interaction takes place. Because angular momentum is conserved during circularization it is actually the distribution of semi-latera recta that is appropriate (Hurley, Tout & Pols 2002).

References

Arnett, W. D. 1982, *ApJ*, **253**, 785.
de Bernardis, P., et al. 2000, *Nature*, **404**, 955.
Branch, D. 1998, *ARA&A*, **36**, 17.
Cappellaro, E., Turatto, M., Tsvetkov, D. Yu., Bartunov, O. S., Pollas, C., Evans, R., Hamuy, M. 1997, *A&A*, **322**, 431.
Eggleton, P. P. 1971, *MNRAS*, **151**, 351.
Hachisu, I., Kato, M., Nomoto, K. 1996, *ApJ*, **470**, L97.
Hamuy, M., Phillips, M., Suntzeff, N., Maza, J. 2003, *IAUC*, **8151**, 2.
Hurley, J. R., Tout, C. A., Pols, O. R. 2002, *MNRAS*, **329**, 897.
Izzard, R. G., Tout, C. A. 2003, *PASA*, **20**, 345.
Kahabka, P., van den Heuvel, E. P. J. 1997, *ARA&A*, **35**, 69.
Kawai, Y., Saio, H., Nomoto, K. 1987, *ApJ*, **315**, 229.
King, A. R., Rolfe, D. J., Schenker, K. 2003, *MNRAS*, **341**, L35.

Kroupa, P., Tout, C. A., Gilmore, G. 1993, *MNRAS*, **262**, 545.
Lesaffre, P., Podsiadlowski, P., Tout, C. A. 2005, *MNRAS*, **356**, 131.
Livio, M., Riess, A. G. 2003, *ApJ*, **594**, L93.
Livne, E., Arnett, D. 1995, *ApJ*, **452**, 62.
Maxted, P. F. L., Marsh, T. R. 1999, *MNRAS*, **307**, 122.
Mazzali, P. A., Lucy, L. B. 1998, *MNRAS*, **295**, 428.
Mazzali, P. A., Nomoto, K., Cappellaro, E., Nakamura, T., Umeda, H., Iwamoto, K. 2001, *ApJ*, **547**, 988.
Nelemans, G., Tout, C. A. 2004, *MNRAS*, **356**, 753.
Nelemans, G., Verbunt, F., Yungelson, L. R., Portegies Zwart, S. F. 2000, *A&A*, **360**, 1011.
Nomoto, K., Iben, I. Jr. 1985, *ApJ*, **297**, 531.
Paczyński, B., Żytkow, A. N. 1978, *ApJ*, **222**, 604.
Perlmutter, S., et al. 1999, *ApJ*, **517**, 565.
Phillips, M. M. 1993, *ApJ*, **413**, L105.
Regős, E., Tout, C. A., Wickramasinghe, D., Hurley, J. R., Pols, O. R. 2003, *NewAst*, **8**, 283.
Riess, A. G., et al. 1998, *AJ*, **116**, 1009.
Röpke, F. K., Hillebrandt, W. 2004, *A&A*, **420**, L1.
Rowan-Robinson, M. 2002, *MNRAS*, **332**, 352.
Thielemann, F.-K., Nomoto, K., Yokoi, K. 1986, *A&A*, **158**, 17.
Timmes, F. X., Brown, E. F., Truran, J. W. 2003, *ApJ*, **590**, L83.
Timmes, F. X., Woosley, S. E., Hartmann, D. H., Hoffman, R. D. 1996, *ApJ*, **464**, 332.
Tout, C. A., Regős, E., Wickramasinghe, D., Hurley, J. R., Pols, O. R. 2001, *MmSAI*, **72**, 37.
Warner, B. 1995, *Cataclysmic Variables* CUP, Cambridge.
Wittman, D. M., Tyson, J. A., Kirkman, D., Dell'Antonio, I., Bernstein, G. 2000, *Nature*, **405**, 143.
Woosley, S. E., Taam, R. E., Weaver, T. A. 1986, *ApJ*, **301**, 601.
Woosley, S. E., Weaver, T. A. 1994, *ApJ*, **423**, 371.
Yoon, S. C., Langer, N. 2004, *A&A*, **419**, 623.

SURVEYS FOR DOUBLE DEGENERATE PROGENITORS OF SUPERNOVAE TYPE IA* † ‡

R. Napiwotzki[1,2], C.A. Karl[2], G. Nelemans[3], L. Yungelson[4], N. Christlieb[5], H. Drechsel[2], U. Heber[2], D. Homeier[6], D. Koester[7], B. Leibundgut[8], T.R. Marsh[9], S. Moehler[7], E.-M. Pauli[2] A. Renzini[8], D. Reimers[5]

[1]*Dept. of Physics & Astronomy, University of Leicester, Leicester, UK,* [2]*Remeis-Sternwarte, Universität Erlangen-Nürnberg, Bamberg, Germany,* [3]*Institute of Astronomy, Cambridge, UK,* [4]*Inst. of Astronomy of the Russian Academy of Sciences, Moscow, Russia,* [5]*Hamburger Sternwarte, Universität Hamburg, Hamburg, Germany,* [6]*Dept. of Physics & Astronomy, University of Georgia, Athens, GA, USA,* [7]*Inst. für Theo. Physik und Astrophysik, Universität Kiel, Kiel, Germany,* [8]*European Southern Observatory, Garching, Germany,* [9]*Department of Physics, University of Warwick, Coventry, UK*

Abstract We report on systematic radial velocity surveys for white dwarf – white dwarf binaries (double degenerates – DDs) including SPY (ESO Supernovae Ia progenitor survey) recently carried out at the VLT. A large sample of DD will allow us to put strong constrains on the phases of close binary evolution of the progenitor systems and to perform an observational test of the DD scenario for supernovae of type Ia. We explain how parameters of the binaries can be derived from various methods. Results for a sample of DDs are presented and discussed.

Keywords: close binaries, double degenerates, supernovae

1. Introduction

Supernovae of type Ia (SN Ia) play an outstanding role for our understanding of galactic evolution and the determination of the extragalactic distance scale. However, the nature of their progenitors is not yet settled (e.g. Livio 2000). According to the current consensus SN Ia explosions happen when white dwarfs

---

*Based on data obtained at the Paranal Observatory of the European Southern Observatory for programs 165.H-0588, 167.D-0407, and 266.D-5658

†Based on observations at the Calar Alto Observatory, Spain which is operated by the Max-Planck-Institute für Astronomie, Heidelberg

‡Based on observations made with the WHT and INT operated on the island of La Palma by the Isaac Newton Group in the Spanish Observatorio del Roque de los Muchachos of the Instituto de Astrofisica de Canarias

E.M. Sion, S. Vennes and H.L. Shipman (eds.), White Dwarfs: Cosmological and Galactic Probes, 153-162.
© 2005 *Springer.*

(WDs) grow to the Chandrasekhar mass of $\approx 1.4 M_{\odot}$. Since no way is known how this can happen to a single WD, this can only be achieved by mass transfer in a binary system.

Several channels have been identified as possibly yielding such a critical mass. They can be broadly grouped into two classes. The single degenerate (SD) channel (Whelan & Iben 1973) in which the WD is accompanied by either a main sequence star, a (super)giant, or a helium star, as mass donor and the double degenerate (DD) channel where the companion is another WD (Webbink 1984; Iben & Tutukov 1984). Close DDs radiate gravitational waves, which results in a shrinking orbit due to the loss of energy and angular momentum. If the initial separation is close enough (orbital periods below ≈ 10 h), a DD system could merge within a Hubble time, and if the combined mass exceeds the Chandrasekhar limit the DD would qualify as a potential SN Ia progenitor.

2. Surveys for close DD

The orbital velocity of WDs in potential SN Ia progenitor systems must be large ($> 150\,\mathrm{km\,s^{-1}}$) making radial velocity (RV) surveys of WDs the most promising detection method. Most WDs are of the hydrogen-rich spectral type DA, displaying broad hydrogen Balmer lines. The remaining WDs are of non-DA spectral types (e.g. DB and DO) and their atmospheres contain no or very little hydrogen. Accurate RV measurements are possible for DA WDs thanks to sharp cores of the Hα profiles caused by NLTE effects.

The first systematic search for DDs among white dwarfs was performed by Robinson & Shafter (1987). They applied a photometric technique with narrow band filters centred on the wings of Hγ or He I 4471Å for DA and DB WDs, respectively. RV velocity variations should produce brightness variations in these filters. This survey investigated 44 WDs, but no RV variable systems were detected. A number of spectroscopic studies (Bragaglia et al. 1990; Foss et al. 1991; Marsh et al. 1995, Saffer et al. 1998; Maxted & Marsh 1999; Maxted et al. 2000a) increased the number of WDs checked for RV variations with sufficient accuracy to 2000. Eighteen DDs with periods $P < 6.3$ d were detected (see Marsh 2000 for a compilation). However, none of these systems seems massive enough to qualify as a SN Ia precursor. This is not surprising, as theoretical simulations suggests that only a few percent of all DDs are potential SN Ia progenitors (Iben, Tutukov, & Yungelson 1997; Nelemans et al. 2001). Note that some of the surveys were even biased against finding SN Ia progenitors, because they focused on low mass WDs. It is obvious that larger samples are needed for statistically significant tests.

Recently, subdwarf B (sdB) stars with WD components have been proposed as potential SNe Ia progenitors by Maxted et al. (2000b), who announced the serendipitous discovery of a massive WD companion of the sdB KPD 1930+2752. If the canonical sdB mass of $0.5M_{\odot}$ is adopted, the mass function yields a minimum total mass of the system in excess of the Chandrasekhar limit. Since this system will merge in less than a Hubble time, this makes KPD 1930+2752 a SN Ia progenitor candidate (although this interpretation has been questioned by Ergma et al. 2001).

3. The SPY project

The surveys mentioned above were performed with 3 to 4 m class telescopes. A significant extension of the sample size without the use of larger telescopes would be difficult due to the limited number of bright WDs. This situation changed after the ESO VLT became available. In order to perform a definitive test of the DD scenario we embarked on a large spectroscopic survey of ≈1000 WDs (ESO **SN** Ia **P**rogenitor surve**Y** – SPY). SPY has overcome the main limitation of all efforts so far to detect DDs that are plausible SN Ia precursors: the samples of surveyed objects were too small.

Spectra were taken with the high-resolution UV-Visual Echelle Spectrograph (UVES) of the UT2 telescope (Kueyen) of the ESO VLT in service mode. Our instrument setup provided nearly complete spectral coverage from 3200 Å to 6650 Å with a resolution $R = 18500$ (0.36 Å at Hα). Due to the nature of the project, two spectra at different, "random" epochs separated by at least one day were observed. We routinely measure RVs with an accuracy of $\approx 2\,\mathrm{km\,s^{-1}}$ or better, therefore running only a very small risk of missing a merger precursor, which have orbital velocities of $150\,\mathrm{km\,s^{-1}}$ or higher. A detailed description of the SPY project can be found in Napiwotzki et al. (2001a).

The large programme has finished at the end of March 2003. A total of 1014 stars were observed. This corresponds to 75% of the known WDs accessible by VLT and brighter than $B = 16.5$ (cf. Figl 1. At this time a second spectrum was still lacking for 242 WDs, but observing time has been granted to complete these observations. Currently we could check 875 stars for RV variations, and detected ≈100 new DDs, 16 are double-lined systems (only 6 were known before). The great advantage of double-lined binaries is that they provide us with a well determined total mass (cf. below). Our sample includes many short period binaries (some examples are discussed below), several with masses closer to the Chandrasekhar limit than any system known before, including one possible SN Ia progenitor candidate (cf. Fig. 4). In addition, we detected 19 RV variable systems with a cool main sequence companion (pre-cataclysmic variables; pre-CVs). Some examples of single-lined and double-lined DDs are shown in Figure 2. Our observations have already increased the DD sample by a factor of seven.

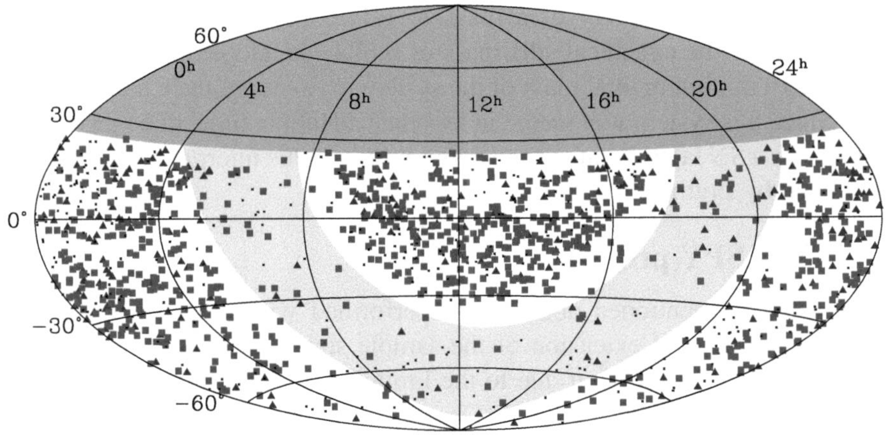

Figure 1. Distribution of all known white dwarfs south of $\delta = +25°$ and brighter than $V = 16.5$. Squares indicate white dwarfs with two spectra taken by SPY. A second spectrum remains to be collected for the triangles, while the dots are the remaining objects without a SPY observation. The light-grey band indicates the position of the Galactic disk ($|b| < 20°$).

Although important information like the periods, which can only be derived from follow-up observations (see below), are presently lacking for most of the stars, the large sample size already allows us to draw some conclusions. (Note that fundamental WD parameters like masses are known from spectral analysis; Koester et al. 2001). One interesting aspect concerns WDs of non-DA classes. Since no sharp NLTE cores are available, non-DA WDs were not included in most RV surveys. SPY is the first RV survey which performs a systematic investigation of both classes of WDs. The use of several helium lines enables us to reach an accuracy similar to the DA case. Our result is that the binary frequency of the non-DA WDs is not significantly different from the value determined for the DA population.

Parameters of double degenerates:. Follow-up observations of this sample are mandatory to exploit its full potential. Periods and WD parameters must be determined to find potential SN Ia progenitors among the candidates. Good statistics of a large DD sample will also set stringent constraints on the evolution of close binaries, which will dramatically improve our understanding of the late stages of their evolution.

The secondary of most DD systems has already cooled down to invisibility. These DDs are single-lined spectroscopic binaries (SB1). Our spectroscopic follow-up observations allow us to determine the orbit of the primary component (i.e. the period P and the RV amplitude K_1). The mass of the primary M_1 is known from a model atmosphere analysis (Koester et al. 2001). Constraints

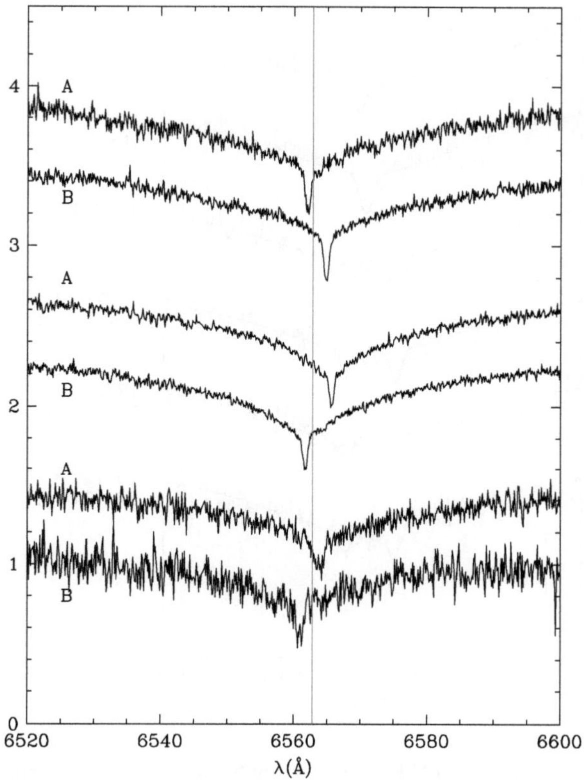

Figure 2. Three single-lined RV variable DDs from our VLT survey. The green line marks the rest wavelength of Hα.

on the mass of the secondary M_2 can be derived from the mass function. For a given inclination angle i the mass of the secondary can be computed. However, i is rarely known, but the result for $i = 90°$ yields a lower mass limit. For a statistical analysis it is useful to adopt the most probable inclination $i = 52°$. We have plotted the single-lined systems with the resulting system mass in Fig. 4. Note that two SB1 binaries have probably combined masses in excess of the Chandrasekhar limit. However, the periods are rather long preventing merging within a few Hubble times.

Sometimes spectral features of both DD components are visible (Fig. 3), i.e. these are double-lined spectroscopic binaries (SB2). As an example for other double-lined systems we discuss here the DA+DA system HE 1414−0848 (Napiwotzki et al. 2002). On one hand the analysis is complicated for double-lined systems, but on the other hand the spectra contain more information than spectra of single-lined systems. The RVs of both WDs can be mea-

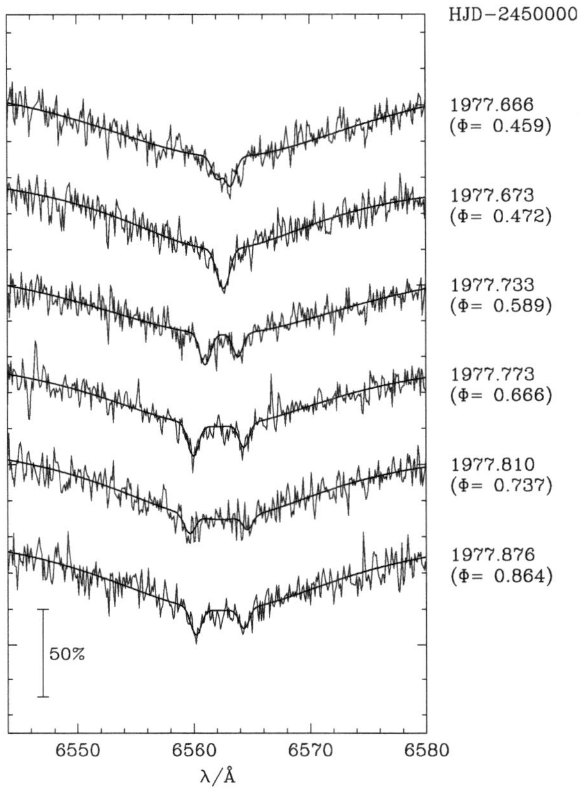

Figure 3. Hα spectra of HE 1414-0848 covering 5 hours during one night together with a fit of the line cores. The numbers indicate the Julian date of the exposures and the orbital phase ϕ. The spectra are slightly rebinned (0.1 Å) without degrading the resolution.

sured, and the orbits of both individual components can be determined (Fig. 5). For our example HE 1414−0848 we derived a period of $P = 12^{\mathrm{h}}25^{\mathrm{m}}44^{\mathrm{s}}$ and semi-amplitudes $K_1 = 127\,\mathrm{km\,s^{-1}}$ and $K_2 = 96\,\mathrm{km\,s^{-1}}$. The ratio of velocity amplitudes is directly related to the mass ratio of both components: $M_2/M_1 = K_1/K_2 = 1.28 \pm 0.02$. However, additional information is needed before the absolute masses can be determined. There exist two options to achieve this goal in double-lined DDs. From Fig. 5 it is evident that the "system velocities" derived for components 1 and 2 differ by $14.3\,\mathrm{km\,s^{-1}}$, which results from the mass dependent gravitational redshift of WDs $z = GM/Rc^2$. This offers the opportunity to determine masses of the individual WDs in double-lined DDs. For a given mass-radius relation gravitational redshifts can be computed as a function of mass. Since the mass ratio is already known from the amplitude radio, only one combination of masses

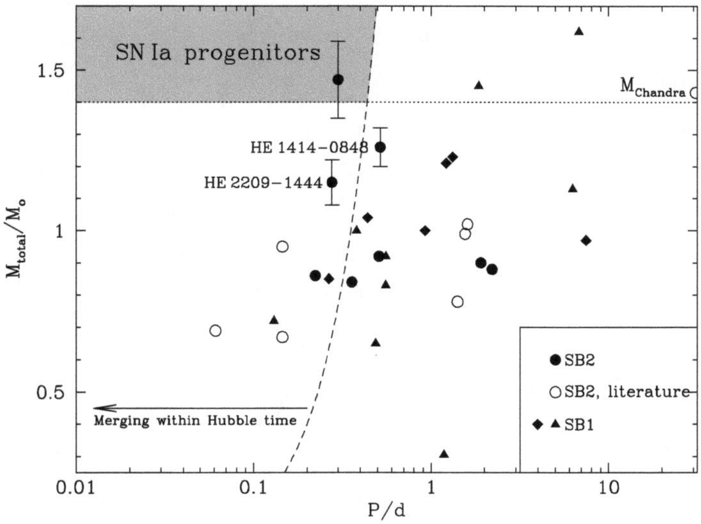

Figure 4. Periods (P) and system masses ($M_{\rm total}$) determined from follow-up observations of DDs from SPY. Results for double-lined systems are compared to previously known systems. The other DD systems are single-lined (triangles: WD primaries; diamonds: sdB primaries). The masses of the unseen companions are estimated from the mass function for the expected average inclination angle ($i = 52°$).

can fulfil both constraints. In the case of HE 1414−0828 we derived individual masses $M_1 = 0.55 \pm 0.03 M_\odot$ and $M_2 = 0.71 \pm 0.03 M_\odot$. The sum of both WD masses is $M = 1.26 \pm 0.06 M_\odot$. Thus HE 1414−0848 is a massive DD with a total mass only 10% below the Chandrasekhar limit.

This method cannot be used if the systems consist of WDs of low mass, for which the individual gravitational redshifts are small, or if their masses are too similar, because the redshift differences are very small and this method cannot be used to determine absolute masses. Another method, which works in these cases as well, are model atmosphere analyses of the spectra to determine the fundamental parameters, effective temperature and surface gravity $g = GM/R^2$, of the stars. Because this system is double-lined the spectra are a superposition of both individual WD spectra. A direct approach would be to disentangle the observed spectra by deconvolution techniques into the spectra of the individual components. Then we could analyse the spectra by fitting synthetic spectra developed for single-lined WDs to the individual line profiles. Such procedures were successfully applied to main sequence double-lined binaries (as discussed elsewhere in these proceedings). However, they have not been tested for WDs, for which the wavelength shifts caused by orbital motions are much smaller than the line widths of the broad Balmer lines. Therefore we choose a different approach for our analysis of double-lined DD systems. We developed the programme FITSB2, which performs a spectral analysis of both

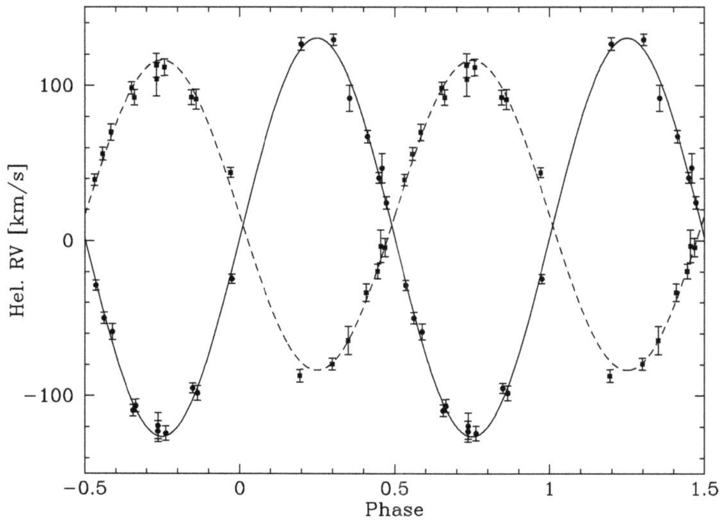

Figure 5. Measured RVs as a function of orbital phase and fitted sine curves for HE 1414−0848. Circles and solid line/rectangles and dashed line indicate the less/more massive component 1/2. Note the difference of the "systemic velocities" γ_0 between both components caused by gravitational redshift.

components of double-lined systems. It is based on a χ^2 minimisation technique using a simplex algorithm. The fit is performed on all available spectra covering different spectral phases simultaneously, i.e. all available spectral information is combined into the parameter determination procedure.

The total number of fit parameters (stellar and orbital) is high. Therefore we fixed as many parameters as possible before performing the model atmosphere analysis. We have kept the RVs of the individual components fixed according to the RV curve. Since the mass ratio is already accurately determined from the RV curve we fixed the gravity ratio. The remaining fit parameters are the effective temperatures of both components and the gravity of the primary. The gravity of the secondary is calculated from that of the primary and the ratios of masses and radii. While the former is known from the analysis of the RV curve, the latter has to be estimated from mass-radius relations. The relative contributions of both stars is determined by their radii and surface fluxes. The flux ratio in the V-band is calculated from the actual parameters and the model fluxes are scaled accordingly. The individual contributions are updated consistently as part of the iteration procedure. The results for HE 1414−0848 are $T_{\rm eff}/\log g$ = 8380 K/7.83 and 10900 K/8.14 for components 1 and 2. A sample fit is shown in Fig. 6. The derived $\log g$ values are in good agreement with the values corresponding to the masses derived from the RV curves: $\log g = 7.92$ and 8.16 respectively.

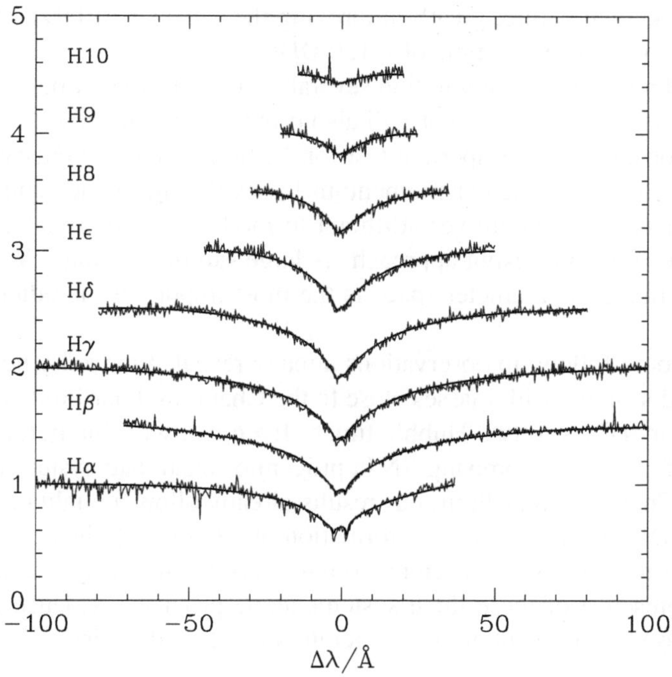

Figure 6. Model atmosphere fit of the Balmer series of HE 1414−0848 with FITSB2. This is only a sample fit. All available spectra, covering different orbital phases, were used simultaneously.

We have plotted HE 1414−0848 as well as our other results on double-lined systems in Fig. 4. Note that one double-lined system is probably a SN Ia progenitor. However, the RV curve of the hotter component is very difficult to measure causing the large error bars. Observing time with the far-UV satellite FUSE has been allocated, which will enable us to measure more accurate RVs. More individual objects are discussed in Napiwotzki et al. (2001b) and Karl et al. (2003).

4. Concluding remarks

The large programme part of SPY has now been completed with some observations underway to complete the observations of the WDs with only one spectrum taken during the survey. We increased the number of WDs checked for RV variability from 200 to 1000 and multiplied the number of known DDs by more than a factor of five (from 18 to ≈120) compared to the results achieved during the last 20 years. Our sample includes many short period binaries (Fig. 4), several with masses closer to the Chandrasekhar limit than any system known before, greatly improving the statistics of DDs. We expect this survey to produce a sample of $\approx$120 DDs.

This will allow us not only to find several of the long sought potential SN Ia precursors (if they are DDs), but will also provide a census of the final binary configurations, hence an important test for the theory of close binary star evolution after mass and angular momentum losses through winds and common envelope phases, which are very difficult to model. An empirical calibration provides the most promising approach. A large sample of binary WDs covering a wide range in parameter space is the most important ingredient for this task.

Our ongoing follow-up observations already revealed the existence of three short period systems with masses close to the Chandrasekhar limit, which will merge within 4 Gyrs to two Hubble times. Even if it will finally turn out that the mass of our most promising SN Ia progenitor candidate system is slightly below the Chandrasekhar limit, our results already allow a qualitative evaluation of the DD channel. Since the formation of a system slightly below Chandrasekhar limit is not very different from the formation of a system above this limit, the presence of these three systems alone provides evidence (although not final proof) that potential DD progenitors of SN Ia do exist.

References

Bragaglia A., Greggio, L., Renzini, A., & D'Odorico, S. 1990, *ApJ*, **365**, L13.

Ergma, E., Fedorova, A.V., & Yungelson, L.R. 2001, *A&A*, **376**, L9.

Foss, D., Wade, R.A., & Green, R.F. 1991, *ApJ*, **374**, 281.

Iben, I.Jr., & Tutukov, A.V. 1984, *ApJS*, **54**, 335.

Iben, I.Jr., Tutukov, A.V., & Yungelson, L.R. 1997, *ApJ*, **475**, 291.

Karl, C.A., Napiwotzki, R., Nelemans, G., et al. 2003, *A&A*, **410**, 663.

Koester, D., Napiwotzki, R., Christlieb, N., et al. 2001, *A&A*, **378**, 556.

Livio, M. 2000 in *Type Ia Supernovae: Theory and Cosmology*, Cambridge Univ. Press, 33.

Marsh, T.R. 2000, *NewAR*, **44**, 119.

Marsh, T.R., Dhillon, V.S., & Duck, S.R. 1995, *MNRAS*, **275**, 828.

Maxted, P.F.L., Marsh, T.R. 1999, *MNRAS*, **307**, 122.

Maxted, P.F.L., Marsh, T.R., & Moran, C.K.J. 2000a, *MNRAS*, **319**, 305.

Maxted, P.F.l., Marsh, T.R., & North, R.C. 2000b, *MNRAS*, **317**, L41.

Napiwotzki, R., Christlieb, N., Drechsel, H., et al. 2001a, *Astronomische Nachrichten*, **322**, 411.

Napiwotzki, R., Edelmann, H., Heber, U., et al. 2001b, *A&A*, **378**, L17.

Napiwotzki, R., Koester, K., Nelemans, G., et al. 2002, *A&A*, **386**, 957.

Nelemans, G., Yungelson, L.R., Portegies Zwart, S.F., & Verbunt, F. 2001, *A&A*, **365**, 491.

Robinson, E.L., & Shafter, A.W. 1987, *ApJ*, **322**, 296.

Saffer, R.A., Livio, M., & Yungelson, L.R. 1998, *ApJ*, **502**, 394.

Webbink, R.F. 1984, *ApJ*, **277**, 355.

Whelan, J., Iben, I.Jr. 1973, *ApJ*, **186**, 1007.

POPULATION SYNTHESIS FOR PROGENITORS OF TYPE IA SUPERNOVAE

Lev R. Yungelson
Institute of Astronomy of the Russian Academy of Sciences
lry@inasan.rssi.ru

Abstract We discuss the application of population synthesis for binary stars to progenitors of SN Ia. We show that the only candidate systems able to support the rate of SNe Ia $\nu_{\rm Ia} \sim 10^{-3}\,{\rm yr}^{-1}$ both in old and young populations are merging white dwarfs. In young populations (~ 1 Gyr) edge-lit detonations in semidetached systems with nondegenerate helium star donors are also able to support a similar $\nu_{\rm Ia}$. The estimated current Galactic rate of SN Ia with single-degenerate progenitors is $\sim 10^{-4}\,{\rm yr}^{-1}$.

Keywords: supernovae

1. Introduction

There is little doubt that explosions of SN Ia are thermonuclear disruptions of the mass-accreting carbon-oxygen white dwarfs (CO WD) in binaries. The main facts arguing for this are: released energy per 1 g is comparable to $\epsilon_{CO \rightarrow Fe}$; explosive nature of the events suggests that degeneracy plays a significant role; explosions may occur long after cessation of star formation; hydrogen is not detected in the spectra of SN Ia (but see discussion of single-degenerate scenario below).

Identification of the SN Ia progenitors is important for several reasons. It may help to constrain the theory of binary–star evolution. Modeling and understanding the explosions requires knowledge of the initial conditions for them and the environments in which they take place. Evolution of the galaxies depends on the radiative, kinetic energy, and nucleosynthetic output of SN Ia and the evolution of SN Ia rate in time, which, in turn, depend on the nature of the progenitor systems. The nature of the progenitors is related to the use of SN Ia as distance indicators for determination of cosmological parameters H_0 and q_0. Evolution of the luminosity function and the rate of SNe is important in this respect.

E.M. Sion, S. Vennes and H.L. Shipman (eds.), White Dwarfs: Cosmological and Galactic Probes, 163-173.
© 2005 *Springer.*

Table 1. Occurrence rates of SNe Ia in candidate progenitor systems (in yr^{-1})

Donor	CO WD	MS/SG	He star	He WD	RG
Counterpart	Close Binary WD	Supersoft XRS	Blue sd	AM CVn	Symbiotic Star
Mass transfer mode	Merger	RLOF	RLOF	RLOF	Wind
	Direct carbon ignition (Chandrasekhar SN)				
Young population	$\mathbf{10^{-3}}$	10^{-4}	10^{-4}	10^{-5}	10^{-6}
Old population	$\mathbf{10^{-3}}$	–	–	10^{-5}	10^{-6}
	Edge-lit detonation (sub-Chandrasekhar SN)				
Young population	–	$\lesssim 10^{-4}$	$\mathbf{10^{-3}}$	–	$\lesssim 10^{-3}$

A successful model for the population of progenitors of SN Ia has to explain the inferred Galactic rate of events $(4 \pm 2) \cdot 10^{-3}\,\mathrm{yr}^{-1}$ (Cappellaro and Turatto 2001), the origin of the observational diversity among local ($z < 0.1$) SNe Ia — $36 \pm 9\%$ may be "peculiar" (Li et al. 2001), and the occurrence of SNe Ia in stellar populations having a wide range of ages.

Below, we discuss the scenarii of formation of binary systems in which SN Ia may occur and the rate of SN Ia, ν_{Ia}, predicted by different scenarii.

2. Population synthesis

The data provided by stellar evolution theory allows to construct numerical evolutionary scenarii that describe the sequence of transformations of a binary system with given initial masses of components and their separation (M_{10}, M_{20}, a_0) that it can experience in its lifetime.

Statistical studies of stars provide information on the binarity rate and the distributions of binaries over M_{10}, a_0, $q_0 = M_{20}/M_{10}$. Combined with star formation history, this allows to estimate the birthrate of the systems with a given set of (M_{10}, M_{20}, a_0) at any epoch. Then, it is possible to compute their contribution to the past or present population of stars of different types. Integration over whole space of initial parameters or Monte Carlo simulation for a large sample of initial "binaries" gives a complete model of the population of binaries and occurrence rates of different events, e. g., SN. Objects of the same type may be formed by several routes, hence, one may expect variations of SN Ia.

3. Evolutionary scenarii for progenitors of SN Ia

Figure 1 shows (not to scale) a simplified flowchart of the main scenarii in which one may expect formation of a progenitor of SN Ia – a CO WD that may ignite carbon in the center. The rates of formation of potential SN Ia via different channels are summarized in Table 1.

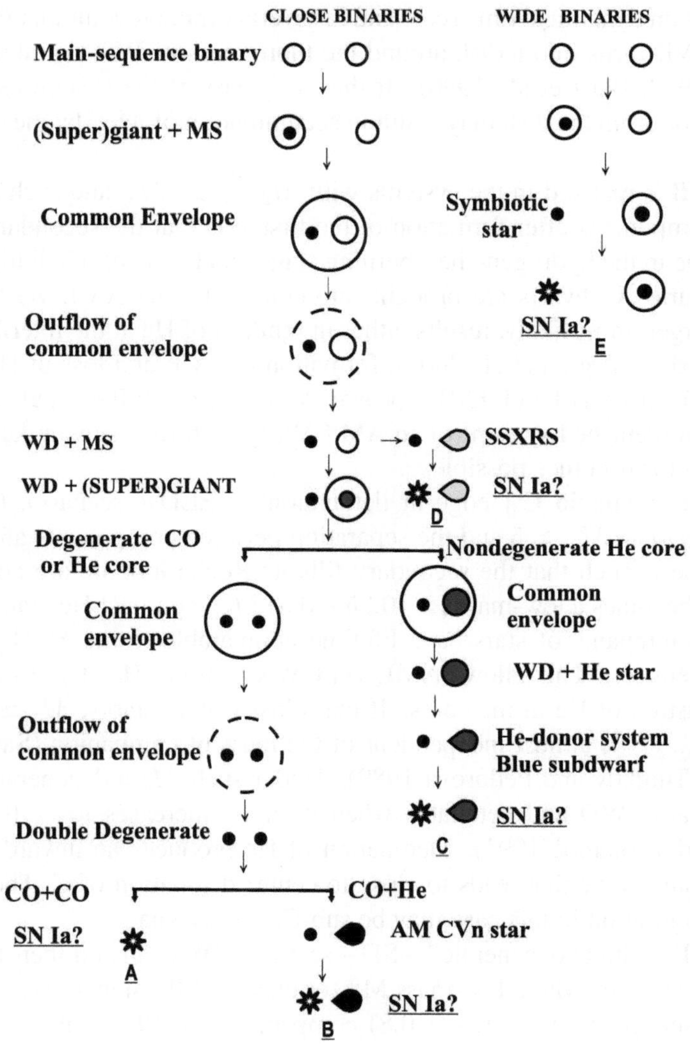

Figure 1. Evolutionary scenarii for possible progenitors of SN Ia.

Scenario A ["double-degenerate"– DD – scenario, (Tutukov and Yungelson 1981; Webbink 1984; Iben and Tutukov 1984)] starts with a main-sequence (MS) binary with $M_{10}, M_{20} \sim (4-10)\, M_{\odot}$. The system is wide enough for Roche-lobe overflow (RLOF) to occur when the primary is an AGB star with a degenerate CO core. After RLOF, a common envelope (CE) forms. If components do not merge inside CE, the core of the primary becomes a CO WD. After dispersal of CE, the system remains wide enough for the secondary to become a CO WD too. The angular momentum loss (AML) via gravitational waves radiation (GWR) results in the RLOF by the lighter of two WD. Mass loss proceeds on dynamical time scale and in several orbital revolutions Roche-lobe filling WD turns into a disk around the more massive WD (Tutukov and Yungelson 1979; Benz et al. 1990). If the total mass of the system exceeds $M_{\rm Ch}$, accretion from the disk may result in accumulation of $M_{\rm Ch}$ by the "core" and SN Ia.

Scenario B is realized in the systems with $M_{20} \lesssim 2.5\,M_{\odot}$ and such a separation of components after formation of the first WD that the secondary fills its Roche lobe in the hydrogen-shell burning stage and becomes a helium WD. Like in scenario **A**, dwarfs are brought into contact by the AML via GWR. Unstable merger, most likely, results either in ignition of He at the interface of accretor and disk (Ergma et al. 2001), formation of a CE and loss of He-rich matter or in formation of an R CrB-type star (Webbink 1984; Iben et al. 1996). If a stable semidetached system (of an AM CVn-type) forms, accumulation of $M_{\rm Ch}$ by accretor becomes possible.

In scenario **Scenario C** ["edge-lit detonation" – ELD – scenario, (Livne 1990)] $2.5 \lesssim M_{20}/M_{\odot} \lesssim 5$ and the separation between components after the first CE phase is such that the secondary fills its Roche lobe before core He ignition and becomes a low-mass [$\simeq (0.35 - 0.8)\,M_{\odot}$] compact He-star. Low-mass helium remnants of stars have lifetime comparable to the MS-lifetime of their progenitors. This allows AML via GWR to bring He-stars to RLOF before exhaustion of He in the cores. If mass loss occurs stably, $\dot{M}_{\rm a} \simeq (2 - 3) \cdot 10^{-8}$ $\rm M_{\odot}\ yr^{-1}$, almost independent of the mass of companion (Savonije et al. 1986; Tutukov and Fedorova 1989). Under such $\dot{M}_{\rm a}$ a degenerate He-layer forms atop WD and detonates when its mass increases to $\sim 0.1\,M_{\odot}$ (Limongi and Tornambè 1991). Detonation of He produces an inward propagating pressure wave that leads to close-to-center detonation of C. The total mass of configuration in this case may be sub-Chandrasekhar.

Scenario D ["single degenerate" – SD – scenario, (Whelan and Iben 1973)] occurs in the systems where low-mass MS (or close to MS) stars ($M_{20} \lesssim (2 - 3)\,M_{\odot}$) or (sub)giant ($M_{20}/M_1 \lesssim 0.8$) companions to WD stably overflow Roche lobes. Accreted hydrogen burns into helium and then into CO-mixture. This allows to accumulate $M_{\rm Ch}$.

Scenario E is the only way to produce SN Ia in a wide system, via accumulation of a He layer for ELD or $M_{\rm Ch}$ by accretion of stellar wind matter in a symbiotic binary (Tutukov and Yungelson 1976).

Scenarii **A** – **E** are associated with binaries of different types and with different masses of components. This sets an "evolutionary clock" – the time delay between formation of a binary and SN Ia. Figure 2 shows the differential rates of SN Ia produced via channels **A**, **C**, and **D** after a burst of star formation. The DD-scenario is the only one that may operate in the populations of

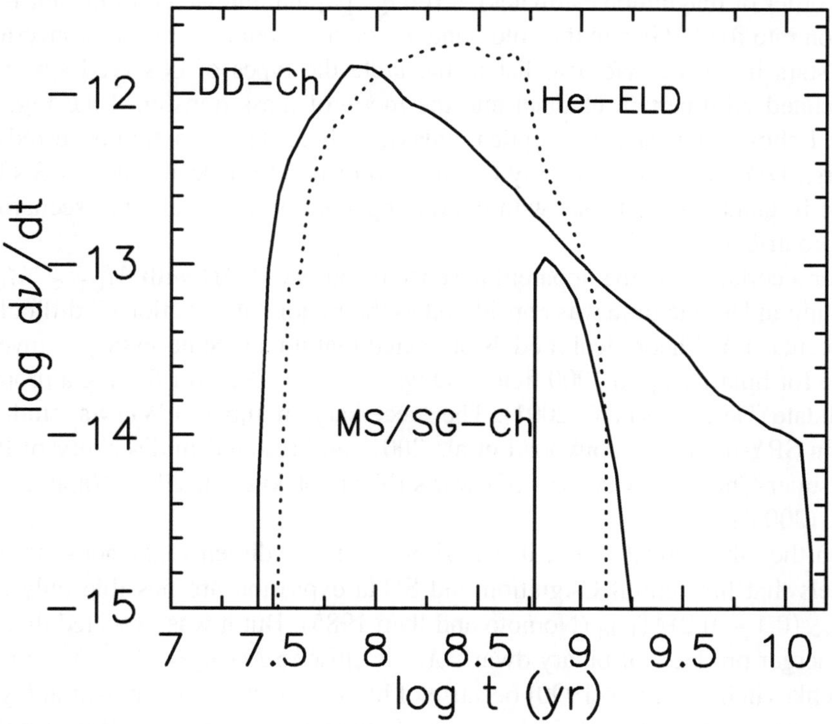

Figure 2. Rates of potential SN Ia-scale events after a 1-yr long star formation burst that produces 1 $M_{\odot}$ of close binary stars.

any age, while SD- or ELD-scenarii are not effective if star formation ceased several Gyr ago.

Table 1 presents the order of magnitude model estimates for $\nu_{\rm Ia}$ after 10 Gyr since beginning of star formation in the populations that have similar total mass comparable to the mass of the Galactic disk. Computations were made by the code used, e. g., by Tutukov and Yungelson (1994) and Yungelson and Livio (1998) for the value of common envelope parameter $\alpha_{ce} = 1$. (Differences in assumptions in population synthesis codes or parameters of computations result in numbers that vary by a factor of several; this is the reason for giving only order of magnitude estimates). "Young" population had constant star formation rate for 10 Gyr; in the "old" one the same amount of gas was converted into stars in 1 Gyr. We also list in the table the types of observed systems associated with certain channel and the mode of mass transfer. Like Fig. 2, table 1 shows that, say, for elliptical galaxies where star formation occurred in a burst, DD-scenario is the only one able to respond for occurrence of SN Ia, while in giant disk galaxies with continuing star formation another scenarios may contribute as well.

For a certain time the apparent absence of observed DD with $M_{\rm tot} \geq M_{\rm Ch}$ merging in Hubble time was considered as the major "observational" difficulty for scenario **A**. Theoretical models predicted that it may be necessary to investigate for binarity up to 1000 field WD with $V \lesssim 16 \div 17$ for finding a proper candidate (Nelemans et al. 2001). The "necessary" number of WD was studied within SPY-project (Napiwotzki et al. 2001) and resulted in discovery of the first super-Chandrasekhar pair of dwarfs [Napiwotzki et al. 2005; Napiwotzki et al. (2003)].

On the "theoretical" side, it was shown for one-dimensional non-rotating models that the central C-ignition and SN Ia explosion are possible only for $\dot{M}_{\rm a} \lesssim (0.1 - 0.2)\dot{M}_{\rm Edd}$ (Nomoto and Iben 1985). But it was expected that in the merger products of binary dwarfs $\dot{M}_{\rm a}$ is close to $\dot{M}_{\rm Edd} \sim 10^{-5}$ $M_\odot$ yr^{-1} (Mochkovitch and Livio 1990) because of high viscosity in the transition layer between the core and the disk. For such $\dot{M}_{\rm a}$ the nuclear burning will start at the core edge, propagate inward and convert the dwarf into an ONeMg one. The latter will collapse without SN Ia (Isern et al. 1983). However, consideration of the role of deposition of angular momentum into central object (Piersanti et al., 2003a,b) has shown that, as a result of spin-up of rotation of WD, instabilities associated with rotation, deformation of WD and angular momentum loss by distorted configuration via GWR, $\dot{M}_{\rm a}$ that is initially $\sim 10^{-5}$ $M_\odot$ yr^{-1}, decreases to $\simeq 4 \cdot 10^{-7}$ $M_\odot$ yr^{-1}. For this $\dot{M}_{\rm a}$ close-to-center ignition of carbon becomes possible.

Because of long apparent absence of an observed "loaded gun" for the DD-scenario and its "theoretical problems", SD-scenario (**D**) is often considered as the most promising one. However, it also encounters severe problems. No

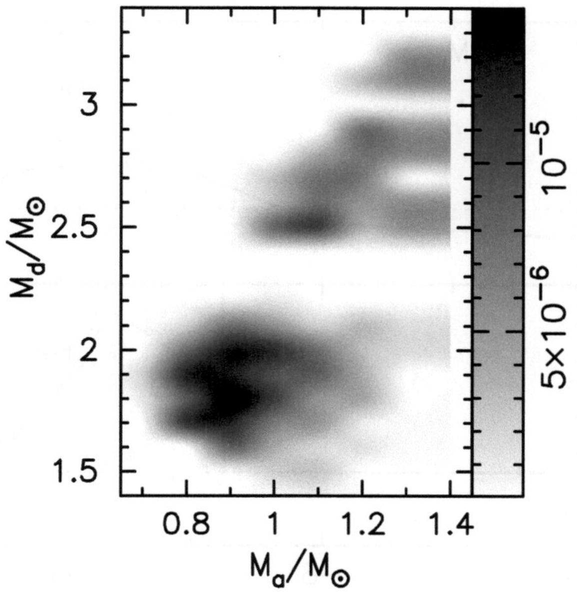

Figure 3. The rate of accumulation of $M_{\rm Ch}$ in the SD-scenario (in yr^{-1}), depending on the masses of WD-accretors and MS- or SG-donors at the beginning of accretion stage.

hydrogen is observed in the spectra of SN Ia, while it is expected that ~ 0.15 $M_\odot$ of H-rich matter may be stripped from the companion by the SN shell (Marietta et al. 2000)[1]. Hydrogen may be discovered both in very early and late optical spectra of SN and in radio- and X-ray ranges (Eck et al. 1995; Marietta et al. 2000; Lentz et al. 2002). As well, no expected (Marietta et al. 2000; Canal et al. 2001; Podsiadlowski 2003) high luminosity and/or high velocity former companions to exploding WD were discovered as yet.

In the SD-scenario, hydrogen first burns into helium and then into C/O mixture. However, two circumstances hamper accumulation of $M_{\rm Ch}$. At $\dot{M}_{\rm a} \lesssim 10^{-8}$ M$_\odot$ yr^{-1} all accumulated mass is lost in Novae explosions (Prialnik and Kovetz 1995). Even if $\dot{M}_{\rm a}$ allows accumulation of He-layer, most of the latter is lost after He-flash (Iben and Tutukov 1996; Cassisi et al. 1998; Piersanti et al. 1999), dynamically or via frictional interaction of binary components with giant-size CE. Thus, the results of computations strongly depend on the assumptions on the amount of mass loss in the nuclear-burning flashes. The flashes become less violent and more effective accumulation of matter may occur if mass is transferred on the rate close to the thermal one (Iben and Tutukov 1984; Yungelson and Livio 1998; Ivanova and Taam 2003). However, this assumption seems to lead to overproduction of supersoft X-ray sources

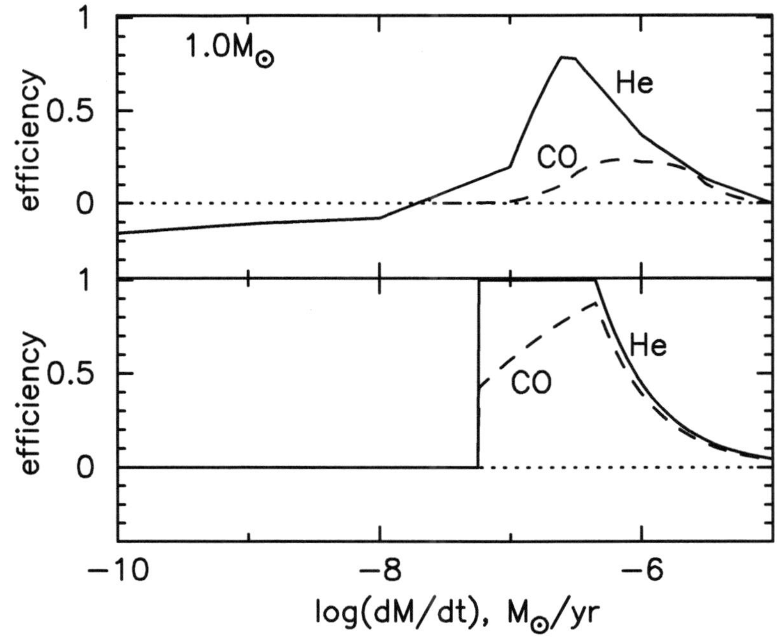

Figure 4. Comparison of efficiency of accumulation of matter by a 1 $M_\odot$ white dwarf under different assumptions. See text for details.

[see the estimate of the number of sources in Fedorova et al. (2004) and completeness of surveys estimates in Di Stefano and (Rappaport 1995)].

The "favorable" range of mass transfer rates widens if mass exchange is stabilized by optically thick stellar wind from WD (Hachisu et al. 1996). Under this assumption (not based on a rigorous treatment of the radiation transfer), the excess of transferred matter over the upper limit for stable hydrogen burning ($\simeq 5 \cdot 10^{-7}$ $\mathrm{M_\odot\ yr^{-1}}$ for a 1 $M_\odot$WD) is blown out of the system taking away specific angular momentum of the WD. This allows to avoid formation of CE for mass transfer rates up to $\simeq 10^{-4}$ $\mathrm{M_\odot\ yr^{-1}}$ and, simultaneously, implies stable hydrogen burning and reduces mass loss in helium burning flashes. Figure 3 shows the range of masses of donors and accretors in "successive" SN Ia progenitors at the beginning of accretion onto the WD stage, obtained under "stabilization" condition and for thermal-time scale mass transfer by Fedorova et al. (2004). The maximum of $\nu_{\rm Ia}$ in the latter study is $2 \cdot 10^{-4}\,\mathrm{yr^{-1}}$, i. e., it still does not exceed $\sim 10\%$ of the inferred Galactic $\nu_{\rm Ia}$. Han & Podsiadlowski (private communication) obtain for this channel the rate up to $1.1 \cdot 10^{-3}\,\mathrm{yr^{-1}}$, closer to the observational estimate.

An important source for discrepant $\nu_{\rm Ia}$ obtained for SD-scenario may be the difference in the assumptions on the mass accumulation efficiency. As an extreme example, the upper panel of Fig. 4 shows the efficiency of accumulation of He and C+O if one takes into account stellar wind mass loss by dwarfs that burn hydrogen steadily, mass loss in Novae explosions after Prialnik and Kovetz (1995) and estimates of mass loss in helium flashes after Iben and Tutukov (1996); the lower panel shows efficiency of accumulation under prescriptions adapted by Han & Podsiadlowski [Fedorova et al. (2004) implemented an "intermediate" case: assumptions on H-accumulation after Prialnik and Kovetz and assumptions on He-burning similar to Han & Podsiadlowski].

Scenario **C** may operate in populations where star formation have ceased no more than $\sim$ 1 Gyr ago and produce SN at the rates that are comparable with the Galactic $\nu_{\rm Ia}$. But the outcome of ELD currently seems to be not compatible with observations of SN Ia. "By construction" of the model, the most rapidly moving products of explosions have to be He and Ni; this is not observed. The spectra produced by ELD are not compatible with observations of the overwhelming majority of SN Ia (Hoeflich et al. 1996). On the theoretical side, it is possible that lifting effect of rotation that reduces effective gravity and degeneracy in the helium layer may prevent detonation (Langer et al. 2003).

Channel **B** most probably gives a very minor contribution to the total SN Ia rate since typical total masses of the systems are well below $M_{\rm Ch}$.

The peculiarity of channel **E** is the behavior of $\dot{M}_{\rm a}$ from the wind: it is initially very low and grows, as companion to the WD expands. Typical initial masses of WD in symbiotic stars are well below 1 $M_{\odot}$ (Yungelson et al. 1995). For them it is more likely to accumulate a helium layer that may be lost in a thermal flash than accumulate $M_{\rm Ch}$.

4. Conclusion

1. Only DD may secure the observed $\nu_{\rm Ia}$ both in old and young populations. Merging pairs with $M_1 + M_2 \simeq M_{\rm Ch}$ were discovered after search in a WD-sample of appropriate size. Account for effects of rotation may solve the problem of central ignition in the merger product. Crucially needed is a study of the physics of merger which follows development of shocks and turbulence in the "transition" zone, transfer of momentum, rotation effects upon evolution of the "core-disk" configuration.

2. Edge-lit detonations in He-accreting systems can be responsible for SN Ia-scale events only in the populations younger than $\sim$ 1 Gyr. Lifting effect of rotation may reduce the number and scale of ELD.

3. Single-degenerate scenario may contribute a fraction ($\sim$ 10%) of all events in young or intermediate age populations. The major obstacle to SD-scenario are H and He thermal flashes. Predictions of the rate of SD-events have to be reconciled with the number of supersoft X-ray sources. A crucial

test for SD-scenario would be detection of H which may be present due to the interaction of SN shell with companion or a "slow wind" of pre-SN.

4. In the DD-scenario one may expect that exploding objects would differ in mass and central C-abundance. In SD-scenario all exploding WD most probably have $M_{\rm Ch}$, but differ in central C. It is unclear whether these differences may explain the diversity of observed SN Ia.

Acknowledgments

This study was supported by RFBR grant 03-02-16254. The author acknowledges financial support of IAU and JD5 SOC that enabled his participation in the meeting.

Notes

1. Recently discovered SN Ia 2001ic and similar 1997cy (Hamuy et al. 2003) may belong to the so-called SN 1.5 type or occur in a symbiotic system (Chugai and Yungelson 2004).

References

Benz, W., Bowers, R. L., Cameron, A. G., and Press, W. H. 1990, *ApJ*, **348**, 647.

Canal, R., Méndez, J., and Ruiz-Lapuente, P. 2001, *ApJL*, **550**, L53.

Cappellaro, E. and Turatto, M. 2001, in *ASSL Vol. 264*, The Influence of Binaries on Stellar Population Studies, 199.

Cassisi, S., Iben, I. Jr., and Tornambe, A. 1998, *ApJ*, **496**, 376.

Chugai, N. and Yungelson, L. 2004, *Astronomy Letters*, **30**, 83 (astro-ph/0308297).

Di Stefano, R. and Rappaport, S. 1995, in *ASP Conf. Ser. 72*, Millisecond Pulsars. A Decade of Surprise, 155.

Eck, C. R., Cowan, J. J., Roberts, D. A., Boffi, F. R., and Branch, D. 1995, *ApJL*, **451**, L53.

Ergma, E., Fedorova, A. V., and Yungelson, L. R. 2001, *A&A*, **376**, L9.

Fedorova, A. V., Tutukov, A. V., and Yungelson, L. R. 2004, *Astronomy Letters*, **30**, 73 (astro-ph/0309052).

Hachisu, I., Kato, M., and Nomoto, K. 1996, *ApJL*, **470**, L97.

Hamuy, M., Phillips, M. M., Suntzeff, N. B., Maza, J., Gonzalez, L. E., Roth, M., Krisciunas, K., Morrell, N., Green, E. M., Persson, S. E., and McCarthy, P. E. 2003, *Nature*, **424**, 651.

Hoeflich, P., Khokhlov, A., Wheeler, J. C., Phillips, M. M., Suntzeff, N. B., and Hamuy, M. 1996, *ApJL*, **472**, L81.

Iben, I. J. and Tutukov, A. V. 1996, *ApJS*, **105**, 145.

Iben, I. Jr. and Tutukov, A. V. 1984, *ApJS*, **54**, 331.

Iben, I. Jr., Tutukov, A. V., and Yungelson, L. R. 1996, *ApJ*, **456**, 750.

Isern, J., Labay, J., Hernanz, M., and Canal, R. 1983, *ApJ*, **273**, 320.

Ivanova, N. and Taam, R. 2003 (astro-ph/0310126v1).

Langer, N., Yoon, S.-C., and Petrovic, J. 2003, In Maeder, A. and Eenens, P., editors, *Stellar Rotation, IAU Symp. 215* (astro-ph/0302232v1).

Lentz, E. J., Baron, E., Hauschildt, P. H., and Branch, D. 2002, *ApJ*, **580**, 374.

Li, W., Filippenko, A. V., Treffers, R. R., Riess, A. G., Hu, J., and Qiu, Y. 2001, *ApJ*, **546**, 734.

Limongi, M. and Tornambè, A. 1991, *ApJ*, **371**, 317.
Livne, E. 1990, *ApJ*, **354**, L53.
Marietta, E., Burrows, A., and Fryxell, B. 2000, *ApJS*, **128**, 615.
Mochkovitch, R. and Livio, M. 1990, *A&A*, **236**, 378.
Napiwotzki, R. et al. 2001, *Nauchnye Informatsii. Ser. Astrof*, **322**, 411.
Napiwotzki, R., et al. 2003, *The Messenger*, **112**, 25.
Napiwotzki, R., et al. 2005, these proceedings.
Nelemans, G., Yungelson, L. R., Portegies Zwart, S. F, and Verbunt, F. 2001, *A&A*, **365**, 491.
Nomoto, K. and Iben, I. Jr. 1985, *ApJ*, **297**, 531.
Piersanti, L., Cassisi, S., Iben, I. Jr., and Tornambé, A. 1999, *ApJL*, **521**, L59.
Piersanti, L., Gagliardi, S., Iben, I. J., and Tornambé, A. 2003a, *ApJ*, **583**, 885.
Piersanti, L., Gagliardi, S., Iben, I. Jr., and Tornambé, A. 2003b, *ApJ*, **598**, 1229.
Podsiadlowski, Ph. 2003 (astro-ph/0303660v1).
Prialnik, D. and Kovetz, A. 1995, *ApJ*, **445**, 789.
Savonije, G. J., de Kool, M., and van den Heuvel, E. P. J. 1986, *A&A*, **155**, 51.
Tutukov, A. V. and Fedorova, A. V. 1989, *Soviet Astr.*, **33**, 606.
Tutukov, A. V. and Yungelson, L. R. 1976, *Astrofizika*, **12**, 521.
Tutukov, A. V. and Yungelson, L. R. 1979, *Acta Astronomica*, **29**, 665.
Tutukov, A. V. and Yungelson, L. R. 1981, *Nauchnye Informatsii. Ser. Astrof*, **49**, 3.
Tutukov, A. V. and Yungelson, L. R. 1994, *MNRAS*, **268**, 871.
Webbink, R. F. 1984, *ApJ*, **277**, 355.
Whelan, J. and Iben, I. Jr. 1973, *ApJ*, **186**, 1007.
Yungelson, L., Livio, M., Tutukov, A., and Kenyon, S. J. 1995, *ApJ*, **447**, 656.
Yungelson, L. R. and Livio, M. 1998, *ApJ*, **497**, 168.

HOW FAR CAN WE TRUST TYPE IA SUPERNOVAE AS STANDARD CANDLES?

Dayal Wickramasinghe[1], Christopher A. Tout[2], Enikő Regős[2,3], Jarrod Hurley[4], Onno Pols [5]
[1]*Australian National University Canberra ACT,* [2]*University of Cambridge,* [3] *Eötvös University, Budapest,* [4]*Monash University,* [5] *University of Utrecht*

Abstract We review the various possibilities that have been proposed as progenitors of Type Ia supernovae (SNe Ia) from the point of view of binary evolution and population synthesis. Depending on the nature of the progenitor, there may be systematic effects that cannot be calibrated by local observations that could undermine their use as standard candles.

Keywords: stellar evolution, close binaries, supernovae, cosmology

1. Introduction

Type Ia supernovae (SNe Ia) are currently being used as standard candles up to red shifts of $z \approx 2$ for constraining cosmological models. SNe Ia near $z \approx 0.5$ provided the first dynamical evidence for a universe whose expansion is accelerating at the current epoch, contrary to expectations from the standard Friedmann models, and led to the re-introduction of the cosmological constant Λ as an essential cosmological parameter (Perlmutter et al. 1999, Riess et al. 1998). A prediction of the Λ cosmologies is that the acceleration should turn over through a coasting phase to a deceleration at earlier epochs as the universe becomes denser and the relative importance of the vacuum energy (provided by the Λ term) decreases. Subsequent observations of even more distant SNe Ia ($z \approx 2$) have shown evidence of deceleration indicating that the turn over occurs at $z \approx 1$ (Riess et al. 2001).

The implications of the SNe Ia results for cosmology are wide ranging but hinge heavily on the standard candle assumption. So what are SNe Ia and is it reasonable to assume that they are standard candles up to red-shift $z \approx 2$ when the universe was younger ($t_{\rm age} \approx \frac{1}{5}t_0 \approx 3\ 10^9$ yr) and had a lower metal abundance?

The major source of energy of a SN Ia is the decay of Ni^{56} to Fe^{56}, and the total energy released is consistent with the decay of approximately one solar

E.M. Sion, S. Vennes and H.L. Shipman (eds.), White Dwarfs: Cosmological and Galactic Probes, 175-189.
© 2005 *Springer.*

mass of Ni^{56}. There are compelling arguments in support of the view that this is related to the thermonuclear detonation of a white dwarf even though the actual explosion mechanism is not fully understood (Hillebrandt and Niemeyer 2000). As a consequence, it has so far not been possible to identify a unique progenitor from theoretical considerations of binary evolution, our understanding of the explosion physics and the observed properties of SNe Ia. According to current thinking such explosions could occur in at least four astrophysical contexts. All involve mass transfer in close interacting binaries containing at least one white dwarf component. Some require Chandrasekhar-mass explosions while in others the white dwarf explodes before the Chandrasekhar mass is reached. It appears likely that there is more than one route via which SNe Ia occur and the observed diversity of their properties even among the normals supports this view (Hamuy et al. 1996).

The empirical relationship between the peak luminosity and light curve speed observed in nearby SNe Ia plays an important role in their use as standard candles because it allows and effective reduction in the standard deviation of absolute luminosities of all SNe Ia to $\pm 15\%$ (Phillips et al. 1999). However, depending on the type of progenitor, these relationships can evolve with metallicity and hence red shift in a manner that cannot be calibrated with observations of local SNe Ia (Regős et al. 2003). Furthermore, *if more than one class is involved*, one may expect changes in the relative mix of the different classes of SNe Ia with redshift. It may not be possible to calibrate such an evolution by local observations; for instance, there may be an evolution of the luminosity function in such a way that the most commonly occurring SNe Ia at high redshifts are not well represented in the local sample.

In this paper we review the likely progenitors of SNe Ia from the point of view of binary evolution and population synthesis calculations and discuss some of the above issues. We link our discussion to calculations carried out with the population synthesis code developed by Hurley et al. (2002).

2. Possible Progenitors of SNe Ia

The lack of hydrogen in the spectra of SNe Ia indicates that the supernova explosion must involve a stellar core that is already devoid of its former hydrogen envelope. Massive stars (above 8 or $9\,M_{\odot}$) are expected to explode before losing their envelopes. In such stars the core collapses to a neutron star or a black hole and the outer regions are likely to be ejected in a type II supernova. Stars of lower mass leave a white dwarf remnant supported by electron degeneracy pressure with masses below the Chandrasekhar limit $M_{\rm Ch}$ with no explosion. These considerations, taken together with the comparative rarity of SNe Ia, have led to the conclusion that their progenitors are interacting binary stars.

The single degenerate hypothesis

Accretion of hydrogen rich material: SG Ch explosions?

The simplest route would be accretion of material on to the white dwarf from a Roche lobe filling companion. Eventually the white dwarf mass could exceed $M_{\rm Ch}$ initiating collapse and, depending on the type of white dwarf, may lead to a SN Ia. Obvious progenitors would be the white dwarfs in the relatively common cataclysmic variables (hereinafter CVs) which are known to be accreting from usually a low-mass main-sequence companion that overfills its Roche lobe (Warner 1995).

The vast majority of the white dwarfs in the CVs are of the CO type. The accreted material is hydrogen rich and is accreted sufficiently slowly ($\dot{M}_2 \leq 10^{-7}\, M_\odot\, {\rm yr}^{-1}$) that a thin degenerate hydrogen layer builds up on the surface. On reaching a mass somewhere between 10^{-5} and $10^{-3}\, M_\odot$ this ignites blowing off as much, or perhaps even more than, has been accreted in a nova explosion. The underlying white dwarf's mass can rarely ever reach $M_{\rm Ch}$.

If the material were to accrete at a somewhat higher rate ($\dot{M} > 10^{-7}\, M_\odot$ ${\rm yr}^{-1}$) the accretion process maintains a surface temperature on the white dwarf sufficient for the accreting hydrogen to burn to helium as it accretes and for the mass of the white dwarf to grow. The white dwarf would then appear as a binary super-soft X-ray source (Kahabka & van den Heuvel 1997). These accretion rates are consistent with mass transfer from a companion more massive than the white dwarf in the shortlived phase of thermal timescale expansion as it evolves from the end of the main sequence to the base of the giant branch, and will normally not lead to the growth of the WD mass up to the Chandrasekhar limit resulting in a SN Ia. For accretion rates above about $3 \times 10^{-7}\, M_\odot\, {\rm yr}^{-1}$ hydrogen rich material accretes too fast to burn immediately and swells up to form a new giant envelope on a thermal timescale which interacts with the companion through the formation of a common envelope (CE) and ends the mass transfer phase.

If mass is transferred from a hydrogen rich star onto a ONe WD which grows in mass to the Chandrasekhar limit, as happens in a subset of the the interacting binaries, the likely end product is an accretion induced collapse into a neutron star and not a SN Ia. If the accretor is a helium white dwarf, it is likely that helium will ignite degenerately when mass builds up to about $0.7 M_\odot$ and detonate the star. We therefore do not expect substantial silicon in the spectra. Such systems are therefore also unlikely to be SNe Ia candidates.

Based on the above discussion, one may be tempted to conclude that the bulk of the SNe Ia are unlikely to be due to Chandrasekhar mass explosions due to the accretion of H rich material form a sub-giant (hereinafter SG-Ch explosions0 However, the situation may not be this simple. Hachisu, Kato & Nomoto (1996, hereinafter HKL) have argued that under certain circumstances a WD that accretes mass from a giant at a high rate ($\dot{M} \geq 4 \times 10^{-7}\ M_\odot\,\mathrm{yr}^{-1}$) may generate a strong optically thick wind that blows off almost all of the transferred matter so that, instead of being forced into CE evolution, the white dwarf simply accretes the H-rich material as it burns to He and grows in mass. If this model is tenable it is possible that *at least some* SNe Ia may result from SG-Ch explosions and others from the degenerate ignition of helium in an Edge-Lit detonation after a critical mass has been accreted (SG-ELDs) (see next section).

Recent support for the above possibility has come from the detection of circumstellar hydrogen in the spectrum of SN 2002ic evidently supporting the view that the companion is a massive AGB star which has lost mass prior to the SN explosion (Hamuy et al. 2003). Furthermore, several recurrent novae with white dwarf components have been observed as supersoft sources (e.g. Kahabka et al. 1999) and analyses of the outburst light curves of recurrent novae such as T CrB, RS Oph, V745 Sco and V3890 Sgr, with allowance for optically thick winds from the white dwarf, have yielded model-dependent white dwarf masses that are very close to the Chandrasekhar limit (Hachishu and Kato 2001a, b). Likewise Hascishu et al. (2000) presented a model of the quiescent accretion disc in U Sco which indicated a white dwarf mass of about $1.37 M_\odot$. However, it is unclear if the underlying white dwarfs in these systems have CO or a ONe compositions. In the latter case, the white dwarf is more likely to collapse into a neutron star than explode as a SN Ia. These results have re-fuelled the suggestion that binary super-soft sources may be the pre-cursors of SNe Ia (HKL, Nomoto et al. 2000) even though the observed frequencies of such events appear to be too low.

Accreting (or merging) CO WDs remain the favoured SN Ia progenitor for the normals largely because of the success of the simplified 1-D delayed-detonation models in explaining the gross chemical and dynamical properties of the ejecta (e.g. Höflich, Wheeler and Thielemann 1998). More realistic 3-D models are currently under construction and it remains to be seen if these are as successful (Gamezo et al. 2003).

Accretion of helium rich material: the edge -lit detonators (ELDs)

Another promising possibility involves the accretion of helium rich material onto a white dwarf. This could occur directly due to mass transfer from a companion, or indirectly due to the burning of hydrogen on the surface of a white dwarf. The material could come from a helium white dwarf that is overflowing its Roche lobe, or from a naked helium star, created either by a very strong wind from a massive star or by the binary stripping of the hydrogen envelope during the red giant evolution of an intermediate-mass star.

Calculations suggest that if the accretor is a helium white dwarf, the white dwarf will detonate through central helium ignition, initiated perhaps by an edge-lit detonation, when the mass grows to about $0.7 M_\odot$. However, burning would not then proceed to the Fe-peak elements and the ejecta may be more appropriate to a SN Ib (Woosley, Taam & Weaver 1986; Nomoto & Sugimoto 1977).

If a CO white dwarf accretes from a He star, the mass ratio is generally small enough for dynamically stable mass transfer. A helium layer builds up on the surface of the CO white dwarf. As in the nova explosions of hydrogen, the base of this helium layer could, under suitable circumstances, reach a temperature at high enough densities for helium to ignite in a degenerate flash and, according to the estimates of Woosley and Weaver (1994), this requires about $0.15 M_\odot$ of helium. It can be envisaged that ignition of such a massive helium layer can detonate the CO core either by compressing it or by an inwardly propagating heating front (Branch and Nomoto 1986; Livne and Glasner 1990). Typical total masses of these edge-lit detonations (hereinafter He ELDs), though below the Chandrasekhar mass, are still of the order of a solar mass. Enough nuclear energy is released to explode the star and produce sufficient Ni^{56} to make a SN Ia.

However, as in the case of accretion of H-rich material, the outcome is expected to depend on the rate of accretion and also on whether the accretion can be approximated to be spherically symmetric. The analysis by Kawai et al. (1987) of helium accreting spherically on to a CO white dwarf showed that at accretion rates above $\dot{M} \approx 3 \times 10^{-8}\, M_\odot\, \mathrm{yr}^{-1}$ the helium burns non-degenerately as it accretes. Models that we have calculated with the Eggleton stellar evolution code (Pols et al., 1995) for spherical accretion confirm this result. They also show that if the accretion rates are less than $\dot{M} \approx 10^{-9}\, M_\odot\, \mathrm{yr}^{-1}$ then degenerate ignition is possible. Our calculations show that, because of their larger size, naked helium stars can feed CO white dwarf at less than this rate but systems with degenerate donors generally do not. A conclusion could therefore be that only the sub-set of the He ELDs where the donor is a naked helium star can yield SNe Ia. However it should be borne in mind that these arguments are based on estimates which assume spherical accretion which may be inappropriate.

The observations of the AM CVn systems may provide some additional insights. The estimated accretion rates in the known AM CVn systems from disc spectra are $6 \times 10^{-9} - 2 \times 10^{-9}\, M_\odot\, \mathrm{yr}^{-1}$ (El-Khoury and Wickramasinghe 2000; Nasser 2001) are typically higher than the limit $\dot{M} \approx 10^{-9}\, M_\odot\, \mathrm{yr}^{-1}$ for degenerate helium burning which would suggest that none of these systems can be SNe Ia. This may be interpreted as evidence that, in the AM CVn systems, accretion occurs from a helium white dwarf donor but when the donor is a naked helium star, a SN Ia occurs (Nelemans et al. 2001).

In this context we should note the properties of V445 Puppis. The absence of hydrogen in the ejecta during the 2000 outburst has led to the suggestion that this is a helium nova caused by a white dwarf accreting from a helium star, a helium white dwarf or a helium-rich main-sequence star (Kato and Hachisu 2003). The mass estimated from a model of the outburst light curves is close to the Chandrasekhar limit. It is unclear whether the outcome will be an accretion induced collapse or a SN Ia (Kato and Hachisu 2003). There could be situations in which He-rich material is transferred on to an ONeWD. The likely outcome is an accretion induced collapse into a neutron star and not a SN Ia.

Accretion of carbon rich material

If carbon rich material is accreted on to a CO or ONeWD, the white dwarf is expected to simply grow in mass until the Chandrasekhar limit is reached. This could lead to a SN Ia depending on the nature of the mass transfer (dynamical or other).

Double Degenerate scenarios: DD Ch explosions

In a very close binary system, gravitational radiation can drive two white dwarfs together until the less massive star fills its Roche lobe. If $q = \frac{M_{\rm donor}}{M_{\rm accretor}} > 0.628$ mass transfer becomes dynamically unstable because a white dwarf expands as it loses mass. In this case a thick accretion disc forms around the more massive white dwarf followed by coalescence on a dynamical time scale.

The merger of two CO WDs could in principle lead to a mass in excess of the Chandrasekhar limit and lead to a SN Ia (hereinafter DD Ch explosions). However, the outcome of such a merger is uncertain. The temperature produced at the core-disc boundary depends on the accretion rate which in turn depends on the viscosity of the disc. If the temperature is hot enough to ignite carbon and oxygen, the WD may be converted to an ONeWD depending on competition between the rate of propagation of the flame inwards which depends on the opacity and the cooling rate of the WD. It is therefore possible that the outcome of a CO+CO WD merger may be an AIC neutron star rather than a SN Ia (Saio & Nomoto 1998). From a theoretical point of view it thus appear that DD Ch explosions are unlikely progenitors of SNe Ia.

If $q = \frac{M_{\rm donor}}{M_{\rm accretor}} < 0.628$ stable mass transfer could occur from the less massive to the more massive white dwarf. However, only a small fraction of CO+CO WD binaries pass through this route and result in Chandrasekhar mass WD explosions. The predicted rates are inadequate to explain the observed SNe Ia rate (see next section).

>From an observational point of view, there has recently been a significant increase in number of new WD+WD binaries through the SPY project. Several have total masses close to the Chandrasekhar limit but only one possible SN Ia candidate has been discovered. Interestingly, within the large uncertainties, the implied merger rate is not inconsistent with what is required to explain SNe Ia (Napiwotzki et al. 2005).

Table 1. Supernovae rates for various progenitors per 1,000 yr.

Metallicity Z	0.02	0.02	0.001	0.01
Common Envelope α	3	1	3	3
C/O + C/O (merger):DD Ch	1.20	0.05	1.63	1.25
C/O on to C/O (accretion)	0.36	0.04	0.55	0.43
H on to C/O (accretion)	0.00	0.00	0.00	0.00
He+C/O (ELD)	8.79	2.90	15.57	10.07
naked He+C/O(ELD)	1.12	0.36	1.35	1.22
Hydrogen + C/O (ELD)	0.00	0.01	0.02	0.00
He wd + He wd (ignition)	1.39	0.15	13.66	2.72
He wd + naked He (ignitiola n)	0.28	0.00	0.06	0.22
Total C/O > MCh	1.57	0.09	2.18	1.69
Total ELD	9.91	3.26	16.95	11.29
Total He wd ignition	1.67	0.15	13.71	2.94

3. Results of Population Synthesis Calculations

The results of our binary population synthesis calculations for different metallicities Z and different values of the common envelope parameter α are presented in Table 1 (see Hurley et al. 2002 for more details). We have distinguished between mergers (dynamically unstable mass transfer) and accretion (stable mass transfer). The results of Table 1 should be be compared with the observed birth rate of SNe Ia for galaxies like our own of $4 \pm 1 \times 10^{-3}\mathrm{yr}^{-1}$ (Cappellaro et al. 1997).

The He-ELDs dominate over CO WD Chandrasekhar mass explosions with the predicted rates exceeding the required value. The naked He Star ELD SNe Ia rate is always smaller that the He WD ELD SNe Ia rate and falls short of the value required to explain the observations. Recall that there are theoretical arguments to suggest that the He WD ELDs may not produce SNe Ia. Likewise the CO on to CO accretion SNe Ia rate is smaller than the CO+CO WD merger SNe Ia rate, and is too small to explain the observed SNe Ia rate. Recall that there are theoretical consideration which suggest that the CO+CO WD mergers will produce AICs and not SN Ia. We note that the absolute supernova rates and the relative importance in the different types both dependent on metallicity. In our calculations, the rates of SNe Ia from hydrogen accretion on to CO WDs are zero because we have not allowed for the HKL optically thick wind loss mechanism.

Yungelson and Livio (2000) have considered the possibility of hydrogen rich donors by allowing for optically thick winds in their population synthesis calculations with a modified version of the HKL model. They find peak rates for SG ELD and SG Ch that are smaller by a factor of about 50 than the CO+CO merger rates.

The above calculations are indicative of likely dominant contributors, but due to the uncertainties of population synthesis calculations, they cannot be used with certainty to exclude any of the possibilities. We expect, however, that the predicted relative changes due to variations in metal abundance represent real trends.

One could in principle turn to the observed properties of the SN Ia to distinguish between the various types but here one is faced with uncertainties in the explosion physics. For instance the observations of the very luminous SN Ia 1991T are consistent with Ni^{56} production both at its centre and in a shell at the outside (Liu et al., 1997 and Fisher et al., 1999). Fisher et al. (1999) favour a Chandrasekhar mass model but the observations appear also to be consistent with the two sites of thermonuclear runaway, the helium envelope and the CO core, present in an ELD. A potential problem with any claim that the majority of SNe Ia are ELDs is the lack of helium found in their spectra (Mazzali and Lucy 1998). However there remains sufficient uncertainty in the explosion models, such as exactly how much helium survives the explosion (typically less than $0.08\,M_\odot$ in the calculations of Livne and Arnett 1995) and the conversion to observed spectra, in particular the assumption that the helium shell remains spherically symmetric around the exploding CO core, that they cannot be ruled out with certainty. A similar problem exists with the single degenerate models with the lack of evidence of hydrogen in the spectra of SNe Ia.

Tout et al. (2001) argued that a thermonuclear runaway that is confined to the CO core (as in CO WD-Ch explosions) or one that occurs both in the CO core and the helium envelopes (as in He ELDs) give different predictions on the production rates of heavier isotopes such as Ti^{44} and Cr^{48} which form preferentially in the envelope. They used the decay product Ca^{44} of Ti^{44}, and the observed ratio of Ca^{44} to Fe^{56} in the solar system to argue that only about 40% of SNe Ia could be edge-lit detonations of CO WDs accreting helium.

A systematic effect

The Phillips relation between peak luminosity and light curve decay speed can be understood (Mazzali et al., 2001) because both the peak luminosity and the light curve speed depend on the amount of Ni^{56} created in the thermonuclear runaway. The more Ni^{56} produced, the brighter the peak while it is the line opacity of the iron peak elements that determines the speed of the light curve, the more iron the slower the light curve. Pinto and Eastman (2001) explain the correlation in terms of more Ni^{56} leading to a hotter, brighter supernova but one with a greater dispersion in the velocity of expanding Ni^{56} layers, which delays the photons because they must escape a wider range of Doppler-shifted transitions.

Let us assume for the moment that the He-ELDs are the sole progenitors of SNe Ia. In ELDs there are two sites of Ni^{56} synthesis. If the peak luminosity is determined primarily by the CO ratio in the core (Höflich et al. 1998) it is primarily a function of the initial main-sequence mass of the progenitor of the CO white dwarf. If the light curve decay speed is determined by the total mass of iron group elements ejected (Pinto and Eastman 2001) this is a function of the total mass of the ELD at the time of explosion because both the CO core and the He envelope are substantially converted to Ni^{56}. In general, binary star evolution ensures that these two masses are correlated and an empirical relation between peak luminosity and light curve shape can be expected.

Regős et al. (2003) used population synthesis calculations for progenitors of different compositions to show that there was indeed such a relationship but that this relationship shifted with metallicity due to the fact that main-sequence stars of a given mass at lower metallicity grow white dwarfs of higher mass (see Figure 1). Their calculations showed that for a given total ELD mass, the main sequence progenitor is typically 30% less massive for a metallicity of $Z = 0.001$ than for a metallicity of $Z = 0.02$. This in turn is expected to result in a systematic shift in the relationship between peak luminosity–light curve speed with red shift in the sense of making distant ELD SNe Ia fainter than those nearby. The ELD rate is predicted to peak between 2×10^8 and 10^9 yr after the progenitors form (Figure 2). The systems we observe today are therefore the product of the appropriate current generation of forming stars. Local supernovae (even those in old galaxies with current low star formation rates) have typically population I progenitors so that local observations cannot be used to calibrate for this effect.

Evolution with metallicity and red shift

Our discussion on ELDs has shown that the intricacies of stellar and binary evolution and the explosion physics could result in systematic changes in the peak luminosity–light curve speed relationship which cannot be calibrated with local SNe Ia. Similar effects may be present for the other possible types of progenitors but may be difficult to unearth until we have a better understanding of the explosion physics. For instance, if SG Ch explosions or SG ELDs are important, we expect the strength of the disc-generated wind in the HKN model to decrease with metallicity. Depending on the explosion physics, this may lead to systematic effects with red shift. If there is a mix of progenitors the relative

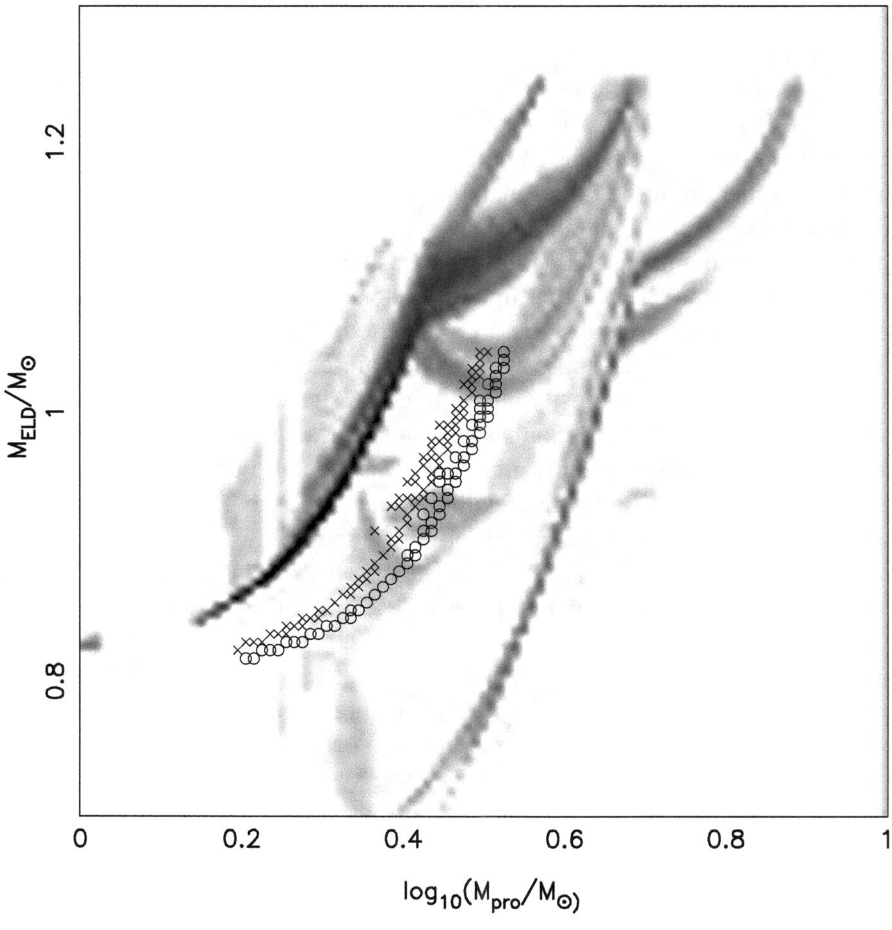

Figure 1. Correlations between total mass of ELDs and the mass of the progenitor of their CO cores evolved from progenitors with $Z = 0.001$. The dark band extending upwards on the left hand side of the diagram consists mainly of He WD ELDs while the narrow band extending upwards on the right hand side mainly of the less common naked He star ELDs. Open circles are the most common ELDs from progenitors of $Z = 0.02$ while crosses are for the most common ELDs from progenitors of $Z = 0.01$.

numbers of the different types can change with red shift and undermine their use as standard candles in a more direct way.

In population synthesis models one deals with a distribution of initial binary parameters which then yield the SNe rates as a function of time. These models show that the SNe event rates rise very sharply after a characteristic time t_s following the birth of the progenitor main-sequence binary systems and declines at different rates thereafter (Figure 2). The time t_s can therefore be used to characterise what is known as the evolutionary clock.

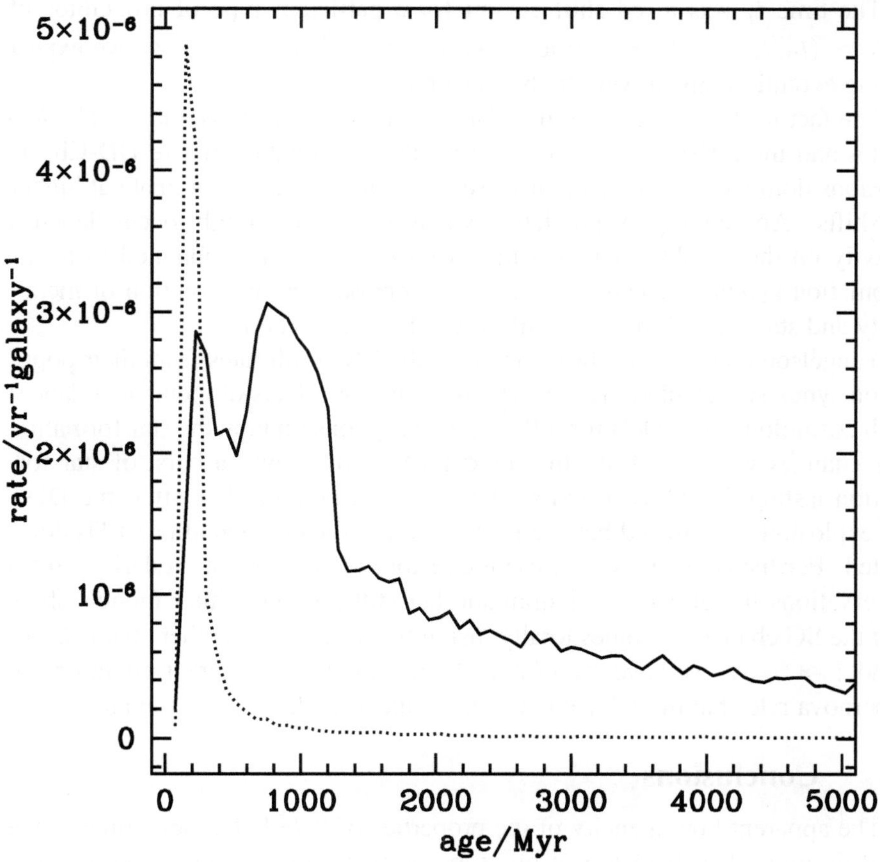

Figure 2. The evolution of birth rate with time since the formation of the binary systems for double degenerate Chandrasekhar mass explosions and He Edge-lit detonations

From our calculations and similar calculations carried out by Yungelson and Livio (2001) we can identify the following general characteristics of the different progenitors

- For DD-Ch $t_s \approx 8 \times 10^7$ yr, and the rate of formation declines by a factor of 100 by $t_f \approx 10^{10}$ yr.

- For naked He star ELDs $t_s \approx 2 \times 10^8$ yr and the rate of formation declines very rapidly, declining by a factor of 100 by $t_f \approx 10^9$yr

- He WD ELDs $t_s \approx 8 \times 10^8$ yr and the rate of formation declines by a factor ≈ 100 by $t_f \approx 10^{10}$ yr.

Yungelson and Livio (2000) find that for SG -Ch explosions $t_s \approx 8 \times 10^8$ yr, but the rate declines very rapidly, reducing by a factor of 100 by $t_f \approx 2 \times 10^9$ yr.

The time t_s sets a red shift cut off for a particular type of progenitor of $z_{\text{crit}} \approx (t_0/t_s)^{\frac{2}{3}} - 1$. For systems with $t_s \approx 10^9$ yr $z_{\text{crit}} \approx 5$ so we expect strong evolution only at very high red shifts.

The factor of 10 difference in t_s for the DD-Ch explosions and the He WD ELDs and the differences in t_f could result in a situation where DD-Ch explosions dominate in the local universe where as the ELDs dominate at higher redshifts. Any attempt at predicting where this switch might occur depends heavily on the absolute numbers that we use for the rates obtained from the population synthesis calculations and assumptions on the evolution of metallicity and star formation history all of which are uncertain.

Yungelson and Livio (2000) convolved the SNe birth rates from their population synthesis calculations (without allowing for effects of metal abundance) with cosmological models for different assumptions on how the star formation rate changes with red shift. In a model where there was a burst of star formation lasting for 4 Gyr at red shift $z = 5$ and nothing thereafter, the DD - Ch explosions dominated between $z \approx 0$ and -0.5 and thereafter ELDs dominated. For their models which use the history of star formation derived from observations the effect is less dramatic but still present. Their models show that the SG channel becomes totally insignificant relative to other channels beyond $z \approx 3$. As we have seen (table 1) metallicity has a direct effect on the supernova rates but these have not been included in the above estimates.

4. Conclusions

The apparent homogeneity of the properties of SNe Ia has sometimes been used to argue that there is just one type of progenitor – namely the detonation of a Chandrasekhar mass CO WD. However, this is no longer supported by observations which show that even among the normals there is a diversity of properties. The possibility of a mix of progenitors therefore appears very likely.

There are at least four main routes of binary evolution that have been identified that could lead to SNe Ia, Chandrasekhar or sub Chandrasekhar mass detonations of a CO WD which accretes from a hydrogen rich donor possibly related to supersoft sources (SG Ch, SG ELD), Chandrasekhar mass explosions due to the merging of two CO WDs on a dynamical time scale (DD Ch), Sub-Chandrasekhar mass detonations of a CO WD that accretes a critical mass of helium from a He WD (He ELD) and Sub-Chandrasekhar mass detonations of a CO WD that accretes a critical mass of helium from a naked He star (Naked

He ELD). Unfortunately there are uncertainties associated with each of these owing to one /or a combination of the following:

- Double degenerate mergers of two CO WDs with a total WD mass that exceeds the Chandrasekhar limit may lead to an accretion induced collapse rather than a SN Ia
- Accretion from a hydrogen-rich sub-giant donor on to a CO WD may not lead to a Chandrasekhar-mass explosion because the mass transfer phase which allows stable nuclear burning is too shortlived unless special mechanisms of mass loss (optically thick winds) are invoked.
- Sub-Chandrasekhar mass detonations of a CO WD which accretes He from a He WD or a naked He star may or may not yield ejecta consistent with observations of the dominant class of SNe Ia (the normals) depending on the uncertain details of the explosion mechanisms.
- The birth rates of the possible progenitors predicted from population synthesis calculations have large uncertainties.
- The birth rates deduced from observations of the space densities of the possible progenitor systems and estimated lifetimes have large uncertainties.

Most models predict that, if there is a mix of progenitors, the relative importance of the different types of progenitor evolves with red shift. Calculations suggest, for instance, that, almost independently of star formation history, the DD-Ch rate dominates over the He ELD rate up to $z \approx 1$ and by $z \approx 2$ the He ELD rate dominates.

Finally, there may be systematic effects on the peak luminosity–decay speed relationship that depend on progenitor composition that cannot be calibrated with local SNe Ia which may undermine their use as distance indicators. We have identified one such possible effect for the ELDs. Similar effects may be present for other progenitors. Until the details of the change from nuclear/thermal to dynamical evolution at ignition are better understood for the various SNe scenarios their use as standard candles to constrain cosmological parameters should not be taken for granted.

Acknowledgments

CAT thanks Churchill College for a Fellowship and DTW acknowledges valuable discussions with Lilia Ferrario.

References

Branch, D. & Nomoto, K., 1986, *A&A*, **164**, L13.

Cappellaro, E., Turatto, M., Tsvetkov, D.Y., Bartunov, O.S., Pollas, C., Evans, R. & Hamuy, M. 1997, *A&A*, **322**, 431.
El-Khoury, W. & Wickramasinghe, D. T. 2000, *A&A*, **358**, 154.
Fisher, A., Branch, D., Hatano, K. & Baron, E. 1999, *MNRAS*, **304**, 67.
Hachisu, I., Kato, M. & Nomoto, K. 1996, *ApJ*, **470**, L97.
Hamuy, M., et al. 1996, *AJ*, **112**, 2398.
Gamezo, V.N., et al. 2003, *Science*, **299**, 77.
Hachisu, I., Kato, M., Kato, M.,Matsumoto, K. & Nomoto, K. 2000, *ApJ*, **534**, L189.
Hachisu, I. & Kato, M. 2001, *ApJ*, **553**, L161.
Hachisu, I. & Kato, M. 2001, *ApJ*, **558**, 323.
Han, Z., Podsiadlowski P. & Eggleton P.P. 1995, *MNRAS*, **272**, 800.
Hillebrandt, W. & Niemeyer, J.C. 2000, *ARA&A*, **38**, 191.
Hurley, J.R., Tout, C.A. & Pols, O.R. 2002, *MNRAS*, **329**, 897.
Höflich, P., Wheeler, J.C. & Thielemann, F.K. 1998, *ApJ*, **495**, 617.
Hamuy, M., et al. 2003, *Nature*, **424**, 651.
Kahabka, P. & van den Heuvel, E.P.J. 1997, *ARA&A*, **35**, 69.
Kahabks, P., Hartmann, H. W., Parma, A. N. & Negueruela, I. 1999, *A&A*, **347**, L43.
Kato, M. & Hachisu, I. 2003, *ApJ*, **598**, L107.
Kawai, Y., Saio, H. & Nomoto, K. 1987, *ApJ*, **315**, 229.
Liu, W., Jeffery, D.J. & Schultz, D.R. 1997, *ApJ*, **486**, L35.
Livne, E. & Arnett, D. 1995, *ApJ*, **452**, 62.
Livne, E. & Glasner, A.S. 1990, *ApJ*, **361**, 244.
Mazzali, P.A. & Lucy, L.B. 1998, *MNRAS*, **295**, 428.
Mazzali, P.A., Nomoto, K., Cappellaro, E., Nakamura, T., Umeda, H. & Iwamoto, K. 2001, *ApJ*, **547**, 988.
Napiwotzki, R., et al. 2005, these proceedings.
Nasser, M. R., Solheim, J. E., & Semionoff, D. A. 2001, *A&A*, **373**, 222.
Nelemans, G., Portegies Zwart, S. F., Verbunt, F. & Yungelson, L. R. 2001, *A&A*, **368**, 939.
Nomoto, K. & Iben Jr., I. 1985, *ApJ*, **297**, 531.
Nomoto, K. & Sugimoto, D. 1977, *PASJ*, **29**, 765.
Nugent, P., Baron, E., Branch, D., Fisher, A. & Hauschildt, P.H. 1997, *ApJ*, **485**, 812.
Paczynski, B., & Zytkow, A. N. 1978, *ApJ*, **222**, 604.
Perlmutter, S., et al. 1999, *ApJ*, **517**, 565.
Phillips, M.M. 1993, *ApJ*, **413**, L105.
Phillips, M.M., Lira, P., Suntzeff, N.B., Schommer, R.A., Hamuy, M. & Maza, J. 1999, *AJ*, **118**, 1766.
Pinto, P.A. & Eastman, R.G. 2001, *New Astronomy*, **6**, 307.
Regős, E., Tout, C. A., Wickramasinghe, D., Hurley, J. R., & Pols, O. R. 2003, *New Astronomy*, **8**, 283.
Riess, A.G., et al. 1998, *AJ*, **116**, 1009.
Riess, A.G., et al. 2001, *ApJ*, **560**, 49.
Saio, H. & Nomoto, K. 1998, *ApJ*, **500**, 388.
Thielemann, F.-K., Nomoto, K. & Yokoi, K. 1986, *A&A*, **158**, 17.
Timmes, F.X., Woosley, S.E., Hartmann, D.H. & Hoffman, R.D. 1996, *ApJ*, **464**, 332.
Tout, C. A., Regős, Wickramasinghe, D. T., Hurley, J. R. & Pols, O. R. 2001, *Memorie della Societa Astronimica Italiana*, **72**, 371.
Umeda, H., Nomoto, K., Kobayashi, C., Hachisu, I. & Kato, M. 1999, *ApJ*, **522**, L43.
Warner, B. 1995, *Cataclysmic Variable Stars*, Cambridge Univ. Press, Cambridge.
Wittman, D.M., Tyson, J.A., Kirkman, D., Dell'Antonio, I. & Bernstein, G. 2000, *Nature*, **405**, 143.

Woosley, S.E., Taam, R.E. & Weaver, T.A. 1986, *ApJ*, **301**, 601.
Woosley, S.E. & Weaver, T.A. 1994, *ApJ*, **423**, 371.
Woosley, S.E. & Weaver, T.A. 1995, *ApJS*, **101**, 181.
Yungelson, L. R. & Livio, M. 2000, *ApJ*, **528**, 108.
Yungelson, L. R. & Livio, M. 1998, *ApJ*, **497**, 168.

KIRSHNER ON WHITE DWARFS, SUPERNOVAE, AND COSMOLOGY

Harry L. Shipman
Department of Physics and Astronomy, University of Delaware, Newark, DE19710, USA

Abstract This paper is a brief report on Bob Kirshner's talk at the conference. He made several principal points which are of interest to scientists who investigate white dwarf stars and binary systems containing white dwarfs. First, a lot of knowledge about the Universe is being hung on Type Ia supernovae. We have no really adequate picture of what stars make them. The class itself may not be a homogeneous class. While no one has definitively identified any physical cause that may make present-day type Ia supernovae different from those that went off 5-7 Gyr ago, there have been suggestions. There may yet be some systematics in the supernova data which will at least question the claimed precision of the current consensus regarding a cosmological model. Most importantly, even if all of the cosmological probes give us the same answer, we should not be satisfied with our current understanding of supernovae. They are important for their own sake.

Keywords: Cosmology, supernovae

1. Introduction

One of the most dramatic developments over the past decade has been the analysis of the Hubble diagram of Type Ia supernovae and its interpretation in terms of an accelerating Universe. This work has come from a number of very large teams of astronomers. At the conference which is the basis of this book, one of the principal figures in those teams, Bob Kirshner, gave an excellent review talk on the current situation. This article is based on my notes of his talk. It has been informed by several recent articles (Tonry et al. 2003; Riess et al. 2004, and especially Kirshner 2003), but I concentrated on the material in the talk and resisted the temptation to write my own review of the literature. Other scientists are much more qualified to do that. Readers who wish to explore the views of particular supernova specialists are encouraged to examine other chapters in this volume, Kirshner's brief but very insightful review, addressed to the general science community, in Science (Kirshner 2003), and Kirshner's very entertaining popular book (Kirshner 2002).

E.M. Sion, S. Vennes and H.L. Shipman (eds.), White Dwarfs: Cosmological and Galactic Probes, 191-194.
© 2005 *Springer*.

2. Supernovae and Cosmology

Ever since the construction of the 200-inch telescope on Mount Palomar about sixty years ago, astronomers have hoped to use the Hubble diagram of apparent magnitude versus redshift to determine the appropriate cosmological model. Much rides on this work. The cosmological model gives us an understanding of the fate of the Universe. The original hope of the builders of the 200-inch was that one could identify some type of galaxy which would be a standard candle, would have the same luminosity everywhere in the Universe, and thus make it relatively easy to fit the Hubble diagram to one of a family of cosmological models. Alas, galaxies could not do the job. In the past decade, Type Ia supernovae have emerged as a possible standard, or at least standardizable, candle.

From redshifts of 0.1 to 0.3, observations of extragalactic supernovae are basically used to understand this class of objects, at least in an empirical way. At these relatively low redshifts, all the cosmological models produce basically the same distance-redshift relation. It was observations of these relatively nearby supernovae that suggested that the Type Ia's had the same luminosity everywhere.

In 2003, the published data, which Kirshner relied on in his talk, showed that the signal of cosmological acceleration (and thus dark energy) came from observations of supernovae which went off between 5-7 Gyr ago, with redshifts greater than 0.3. The effect of the accelerating Universe at these epochs is still relatively modest. There is an 0.2 magnitude difference in the Hubble diagram between a Universe which is accelerating and the reference model, namely the flat, barely bound Universe with (omega) = 1, the popular model of the late 1980s and early 1990s. The most straightforward interpretation of these data is that 70 Hubble constant is 72 km/s/Mpc. The properties of the dark energy are, for the most part, poorly understood. The cosmologically relevant property is that it has a very peculiar equation of state in that the pressure produced by the dark energy is inversely dependent on the density. As the Universe expands, the dark energy density goes down, and so its pressure goes up. This effect becomes cosmologically significant only when the Universe is several billion years old. There are a number of possible systematic errors in this interpretation of the data, systematic errors which will not go away as the sample of supernovae in the 0.3-0.7 range becomes larger. Observers have put the greatest effort into using the shapes of light curves and the morphology of the host galaxy as a way of calibrating the differences in the luminosities of individual type Ia supernovae. Tonry et al. (2003) provide a fairly extensive description of the various schemes (Kirshner's word) that have been used to make the data more homogeneous. These schemes do have the effect of reducing the scatter in the Hubble diagram. However, the class of observed Type

Ia supernovae may still be rather heterogeneous, giving systematic errors the opportunity to hide beneath the surface.

There may be some subtle chemical evolution effects, either produced by an increasing concentration of heavy elements in the gas that forms the stars that make the supernovae, or in the exploding white dwarf itself. One of the issues that repeatedly comes up is that we do not really understand what kinds of stars produce supernovae. To be sure, the model of a white dwarf in a binary system accreting enough material, being pushed over the Chandresekhar limit, and collapsing, has been around for long enough so that it has reached the status of current conventional wisdom. However, there is a shortage of observed systems which could become type Ia supernovae on a reasonable time scale. Other papers presented at the conference (see, e.g., Napiwotzki et al. 2005, Yungelson 2005) describe the observational and theoretical situation. Individual readers can make their own judgement as to how much wiggle room there is. Kirshner's view, which is similar to mine, is that we are far from a point where we could seriously claim to understand where Type Ia supernovae come from.

Somewhat surprisingly (to me at least), the morphology of the galaxy which is a host for the Type Ia supernovae seems to have little correlation with the luminosity of the Type Ia supernovae that result. For example, spirals contain some extra bright supernovae, and some extra faint ones. Ellipticals also contain some extra bright supernovae, and some extra faint ones. The average value of the supernova luminosity was the same in spirals as in ellipticals. But no one in the supernova community is taking the step of arguing that this absence of a correlation means the absence of an effect of the composition of the gas that forms stars with the luminosity of the supernovae that result. We still have too little information.

Between the date of the conference and the time this article was prepared (October 2004), some papers have appeared that begin to probe supernovae in the important cosmological era surrounding $z \sim 1$ (Kirshner 2003; Riess et al. 2004). This epoch is significant because the effects of acceleration in the Hubble diagram change sign. We are now looking back at an epoch before the significant acceleration took place. The quantitative cosmological models show that what happens is that a standard candle, instead of being 0.2 magnitudes brighter than the reference model of a flat, critical-density universe, becomes fainter than the reference model. These data may make it harder to use some peculiar effect produced by the chemical properties or evolution of white dwarf stars to explain away the accelerating Universe.

But no matter. The most important thing, in my mind, that Kirshner said was so well articulated that I can quote it exactly. "Even if all the schemes give us the same answer, concordance, we should not be satisfied until we really understand where the supernovae come from." When I ponder this remark, I find that it is at the same time both obvious and profound. Astronomers seek to understand the Universe, and type Ia supernovae are an important part of that Universe. Consequently its obvious that we should strive to understand where they come from and how they work. But its also profound. Some "precision cosmologists" filled some other meeting rooms in Sydney with claims of understanding the cosmological model to two decimal places. They tended to downplay our interest in the origin and physics of Type Ia supernovae. Even if they are right about the cosmology, they are wrong about the importance of understanding supernovae. Even if understanding where Type Ia supernovae come from turns out to make no difference to the accepted cosmological model, it still is important to understand these enigmatic and important objects.

Acknowledgments

I thank Bob Kirshner, Judi Provencal, Ed Sion, Paula Szkody, and Dayal Wickramasinghe for useful insights on this topic. My astronomical work has been supported by various NASA Guest Investigator programs for the Hubble Space Telescope and the Far Ultraviolet Spectroscopic Explorer, and my scholarly work integrating science research and science education research is supported by the National Science Foundation's Distinguished Teaching Scholars Program (DUE-0306557).

References

Kirshner, R.P. 2002, *The Extravagant Universe: Exploding Stars, Dark Energy, and the Accelerating Cosmos*, (Princeton, NJ: Princeton University Press).
Kirshner, R.P. 2003, *Science*, **300**, 1914.
Napiwotzki, R., et al. 2005, these proceedings.
Riess, A.G., et al. 2004, *ApJ*, **607**, 665.
Tonry, J., et al. 2003, *ApJ*, **594**, 1.
Yungelson, L.R. 2005, these proceedings.

SECTION III. CATACLYSMIC VARIABLES AND WHITE DWARF ACCRETION

ACCRETION IN CATACLYSMIC VARIABLE STARS

Review paper

Brian Warner and Patrick A. Woudt

Department of Astronomy, University of Cape Town, Rondebosch 7700, South Africa
warner@physci.uct.ac.za, pwoudt@circinus.ast.uct.ac.za

Abstract We consider accretion onto the white dwarfs in cataclysmic variables in relation to nova eruptions, dwarf nova outbursts, hibernation and non-radial oscillations.

Keywords: cataclysmic variables, close binaries, photometry

The evolution of the surface temperature of an accreting white dwarf (WD) in a cataclysmic variable (CV) resembles a roller coaster. While still detached the WD cools like a single star, but as soon as mass transfer starts there are episodes of heating and cooling as either high mass transfer ($\dot{M}$) occurs through a stable accretion disc (i.e., as a nova-like variable) or episodically in a thermally unstable accretion disc (i.e., as a dwarf nova); and throughout its life the CV intermittently undergoes heating during the unstable thermonuclear burning that fuels nova eruptions.

Although the duty cycles of dwarf novae are directly observable, those of novae are in general not (the exception being recurrent novae, which is some extremum of a continuous distribution of nova recurrence times). There has been a lack of agreement between the theoretical mass ΔM ejected in a nova eruption and the observed mass of nova shells. Until recently, the most elaborate hydrodynamic and thermonuclear models gave $\Delta M \sim 2 \times 10^{-5}$ M$_\odot$, whereas observational estimates are $\sim 2 \times 10^{-4}$ M$_\odot$. This discrepancy has now probably been removed by the more careful consideration of the thermodynamic structure of accreting white dwarfs (Townsley & Bildsten 2003). Recalculation of CVs through entire nova cycles (as in Shara, Prialnik & Kovetz

E.M. Sion, S. Vennes and H.L. Shipman (eds.), White Dwarfs: Cosmological and Galactic Probes, 197-203.
© 2005 *Springer*.

1993) may now give nova recurrence times ($= \Delta M / < \dot{M} >$) as functions of $< \dot{M} >$ and M_{wd} which are closer to reality.

However, the estimate of the mean value $< \dot{M} >$ between nova eruptions remains difficult. The great majority of novae are seen to erupt from a high $\dot{M}$ state and return to the same state for at least 100 – 200 years after eruption (e.g. Warner 2002). Even though both mass and angular momentum are lost during an eruption, simple conservation laws require that the enhanced $\dot{M}$ that occurs through post-eruptive irradiation of the companion by the heated WD be balanced by a lengthy stage of low $\dot{M}$ (Kovetz, Prialnik & Shara 1988). Whether this is enough to drive the CV into a state of hibernation, with low or zero $\dot{M}$ for great lengths of time, is disputed (e.g. Naylor 2002); it certainly does not seem to happen within the first couple of centuries after eruption, but the complete lack of success in finding the remnants of bright novae recorded in Oriental records of two or more millennia ago (Shara 1989), despite the fact that most modern naked eye novae end up as $m_v \sim 12 - 15$ remnants that are easily recognisable, provides the strongest evidence that hibernation on time scales $\sim 10^3$ y does indeed happen. Until indirect ways of estimating the duty cycle of this process are found, the value of $< \dot{M} >$ will remain very uncertain.

Hibernating novae will appear as detached WD/M dwarf binaries in which the M dwarf almost fills its Roche lobe. Such detached systems are also responsible for the orbital period gap, according to the disrupted magnetic braking model (e.g. Verbunt 1984), and may not be distinguishable from those only temporarily at low $\dot{M}$ because of a nova eruption. Outside the period gap, however, there should be no ambiguity, and it is therefore of interest that there are two CVs that have recently been recognised as hibernating systems – namely BPM 71214, with $P_{orb} = 4.84$ h, and EC 13471$-$1258 with $P_{orb} = 3.62$ d (Kawka & Vennes 2003, 2005; O'Donoghue et al. 2003). These are both relatively bright objects (m_v = 13.6 and 14.8, respectively) and would be among the brightest nova-like variables in the sky if they currently had $\dot{M} \sim 10^{-8}$ M$_\odot$ y^{-1}. Another object, LTT 329 at m_v = 14.5, with weak emission indicative of extremely low $\dot{M}$, has been known for some time (Bragaglia et al. 1990). None of these systems are close to the positions of ancient observed novae (Stephenson 1986).

Typical values of M_V before and after eruption imply $\dot{M} \sim 10^{-8}$ M$_\odot$ y^{-1}, which is an order of magnitude larger than what can be accounted for by magnetic braking alone; these are maintained for at least 200 y after eruption, but we do not know for how long these high rates may operate before eruption. The similar M_V, and hence $\dot{M}$, pre- and asymptotically post-eruption has been explained as an irradiative feedback effect that generates an equilibrium high $\dot{M}$ that prevents the WD from cooling below $\sim$ 50 000 K (Warner 2002). If there are $\sim$ 500 y of $\dot{M}$ at $\sim$ 10 times that dictated by magnetic braking (or GR angular momentum loss), then $\sim$ 5000 y of zero or very low $\dot{M}$ are required to redress the period of high living. These indicate the expected ratio of space densities of such systems, but such large numbers of low $\dot{M}$ systems were not found in an initial search for hibernating CVs (Shara 1989).

However, the systematic discovery of fainter CVs (which is where the very low $\dot{M}$ systems inevitably will be found) has begun in the output of the Sloan Digital Sky Survey, the first two releases of which are now available (Szkody et al. 2002a, 2003). Based on the spectra obtained for these $\sim$ 50 new CVs, about 15% have such low $\dot{M}$ values that the WD is easily detected in the visible spectrum. This is the type of survey that is needed to find very low $\dot{M}$ systems – at least for CVs with short orbital periods where both the WD and its companion are intrinsically faint (very low $\dot{M}$ systems of longer P_{orb} may be harder to recognise in the initial colour survey because they will look like ordinary K or M red dwarfs). With more complete surveys to even fainter limits it should become possible to estimate the true frequency distribution of $\dot{M}$, from which the $\dot{M}$ duty cycle between nova eruptions will follow.

Sion (2003) has shown that the observed surface temperatures T_{eff} of the WDs in dwarf novae can be fully accounted for by the compressional heating that accompanies accretion. For the short orbital periods (i.e., below the orbital period gap) T_{eff} clusters around 15 000 K, which is close to what is predicted for GR-driven evolution. This result depends somewhat on the adopted WD masses. One importance of this result is that T_{eff} is an indirect way of learning something about $< \dot{M} >$ and the $\dot{M}$ duty cycle for dwarf novae.

The observed clustering of T_{eff} around 15 000 K, seen in the list given in Table 1 (based on Sion 2003), contains an observational bias – the stars on which it is based have almost all been found from dwarf nova outbursts, which have intervals that become very long for very low $\dot{M}$. For example, in Table 1 the lowest observed temperatures are correlated with the greatest intervals, T_{out}, between outbursts[1]. There should be other, lower T_{eff} systems, with outburst intervals so long that they are unlikely to have been found via outbursts and are therefore absent from studies made hitherto. But these are the intrinsically faint CVs that are beginning to be found spectroscopically in the Sloan Survey.

This point appears most strongly in the known CV WD primaries that have non-radial oscillations. The ZZ Cet instability strip for isolated WDs lies approximately in the range 11 000 – 12 000 K. It may be modified to some extent in accreting WDs where the outer envelope has a different physical and chemical structure (see, e.g., Townsley & Bildsten 2003). Until recently the only known CV/ZZ combination was GW Lib[2] (Warner & van Zyl 1998), which has T_{eff} = 14 700 K according to Szkody et al. (2002b), indicating that the instability strip may be displaced blueward of that for isolated WDs. Note, however, that there are several dwarf novae in Table 1 that have measured T_{eff}

similar to that of GW Lib, have detectable WD absorption lines in their spectra, and have been sufficiently observed photometrically in quiescence for ZZ Cet type oscillations to have been detected – without success. But they typically have outburst intervals $\sim$ 1 y, which may be too short a time in quiescence for oscillations to grow in[3]. GW Lib, on the other hand, has had only one known outburst (in 1983: by which it was identified as a CV); this incidentally demonstrates that oscillations can appear within less than a decade after outburst.

The next CV/ZZ to be discovered was SDSS 1610 (Woudt & Warner 2004), which was identified in the Sloan Survey first release as a very low $\dot{M}$ system, and has had no known outbursts. On the basis of these two examples alone we are led to suspect that CV/ZZ stars are extremely low $\dot{M}$ systems and their T_{eff} will be found to be lower than that currently accepted for GW Lib (and which in any case may have been overestimated).

Once well calibrated, the relative frequency of CV/ZZ systems will provide a valuable indicator of the number of CVs in the accreting WD instability strip, which is another means of studying the $\dot{M}$ history of CVs. The first Sloan Survey release had just one CV/ZZ (SDSS 1610) out of 25 objects. The second release (Szkody et al. 2003) has four or five potential candidates out of 35 candidates, based on the visibility of the WD absorption spectra, but with only two of these having spectra very closely similar to GW Lib or SDSS 1610 (i.e., rather than like the non-oscillating systems such as Z Cha and OY Car, which have more emission- filled absorption lines). Our observations (Warner & Woudt 2003) nevertheless show that SDSS 0131 and SDSS 2205 are certainly CV/ZZ stars – the fifth candidate has yet to be observed. A frequency $\sim$ 8% among faint CVs is therefore indicated, which will give $\sim$ 32 systems once the estimated 400 new CVs expected in the Sloan Survey (Szkody et al. 2002a) have been found and interrogated photometrically.

The four known CV/ZZ stars have hydrogen emission cores superimposed on the WD absorption lines. For somewhat lower $\dot{M}$ the emission lines would not be so readily observable. It would be worth searching carefully for weak emission cores in known ZZ Cet stars – there could be a hibernating CV hidden among them.

Table 1 also includes T_{eff} measurements for polars, i.e., for the strongly magnetic WDs in CVs. Several of these have lower temperatures than any seen in dwarf novae. Some have T_{eff} that could put them in the instability strip. Most of these are the brightest and best-observed polars, but none have been found to have ZZ Cet behaviour. Theoretical investigations are needed that include the effects of strong fields, which will presumably be found to prevent non-radial oscillations above some critical field strength.

The low T_{eff} in the longer P_{orb} polar RXJ1313 implies a much lower $< \dot{M} >$ than is the case for the commonly observed non-magnetic CVs at

Table 1. White Dwarf temperatures in selected CVs.

Star	P_{orb} *(min)*	T_{eff} *(K)*	T_{out} *(d)*	*References*
Non-magnetic				
GW Lib	76.8	13 300	>7000	Szkody et al. 2002b
BW Scl	78.2	14 800		Szkody et al. 2002c
LL And	79.8	14 300	5000:	Howell et al. 2002a
AL Com	81.6	16 300	325	Szkody et al. 2003
WZ Sge	81.6	15 000	10000:	Sion et al. 1995a
SW UMa	81.8	14 000	954	Gaensicke & Koester 1999
HV Vir	83.5	13 300	3500:	Szkody et al. 2002c
WX Cet	84.0	13 000	1000:	Sion et al. 2003
EG Cnc	86.3	12 300	7000:	Szkody et al. 2002c
BC UMa	90.1	15 200	1500:	Szkody et al. 2002c
EK TrA	90.5	18 800	230	Gaensicke et al. 2001
VY Aqr	90.8	13 500	500:	Sion et al. 2003
OY Car	90.9	16 000	160	Horne et al. 1994
HT Cas	106.1	15 500	166:	Wood et al. 1992
VW Hyi	107.0	22 000	28	Sion et al. 1995b
Z Cha	107.3	15 700	51	Robinson et al. 1995
CU Vel	113.0	15 000	165	Gaensicke & Koester 1999
EF Peg	123	16 600	250:	Howell et al. 2002a
Polars				
EF Eri	81.0	9 500		Howell et al. 2002b
DP Leo	89.8	13 500		Schwope et al. 2002
VV Pup	100.4	9 000		Skzody et al. 1983
V834 Cen	101.5	15 000		Beuremann et al. 1990
BL Hyi	113.7	20 000		Wickramasinghe et al. 1984
ST LMi	113.9	11 000		Mukai & Charles 1986
MR Ser	113.6	<8 500		Schwope & Beuremann 1993
AN UMa	114.8	<20 000		Sion 1991
HU Aqr	125.0	<13 000		Glenn et al. 1994
UZ For	126.5	11 000		Bailey & Cropper 1991
AM Her	185.7	~20 000		De Pasquale & Sion 2001
RXJ1313	251.4	15 000		Gaensicke et al. 2000

: Uncertain value

that P_{orb}. Again there may be an observational bias in action – polars even of low $\dot{M}$ are easily found through their hard X-Ray emission, whereas the comparative rarity of low $\dot{M}$ dwarf novae near P_{orb} of 4 h is probably caused by the effect of irradiation-enhanced $\dot{M}$, which turns them either into high $\dot{M}$ nova-likes, or into extremely low $\dot{M}$ dwarf novae (or even deeply hibernating, essentially zero $\dot{M}$ systems, as in the BPM and EC objects discussed above) that are hard to find (Wu, Wickramasinghe & Warner 1995).

Finally, we draw attention to the AM CVn (helium-transferring CV) systems, where the WD primaries could in principle show non-radial oscillations if they are in the equivalent of the DB variable instability strip. These would have to be found among AM CVn stars that have an appropriate $\dot{M}$ – but that is in just the range where these systems show VY Scl behaviour, and in the high state the accretion luminosity will overwhelm any intrinsic pulsations of the primary, while in the low state $\dot{M}$ is probably too variable to give the oscillations time to grow.

Acknowledgments

BW's research is funded by the University of Cape Town; PAW's research is funded by a strategic grant from the University to BW and by funds from the National Research Foundation.

Notes

1. WZ Sge appears not to fit this correlation, but it has a magnetic primary, which is probably the reason for the large T_{out} (e.g. Warner, Livio & Tout 1992).
2. This ZZ Cet star is overlooked in the total given by Fontaine et al. (2002) and Bergeron et al. (2003).
3. Growth times for non-radial oscillations in ZZ Cet stars can be anywhere from hours to thousands of years, according to which mode is being excited (Goldreich & Wu 1999).

References

Bailey, J. & Cropper, M. 1991, *MNRAS*, **253**, 27.
Bergeron, P., Fontaine, G., Billeres, M., Boudreault, S. & Green, E.M. 2003, *ApJ*, **600**, 404.
Beuermann, K., Schwope, A.D., Thomas, H.-C. & Jordan, S. 1990, In Accretion Powered Compact Binaries, ed. C.W. Mauche, Cambridge University Press, 265.
Bragaglia, A., Greggio, L., Renzini, A. & D'Odorico, S. 1990, *ApJL*, **365**, 13.
De Pasquale, J. & Sion, E.M. 2001, *ApJ*, **557**, 978.
Fontaine, G., Bergeron, P, Billères, M. & Charpinet, S. 2003, *ApJ*, **591**, 1184.
Gaensicke, B.T, & Koester, D. 1999, *A&A*, **346**, 151.
Gaensicke, B.T., Beuermann, K., De Martino, D. & Thomas, H.-C. 2000, *A&A*, **354**, 605.
Gaensicke, B.T., Szkody, P., Sion, E.M., Hoard, D.W., Howell, S., Cheng, F.H. & Hubeny, I. 2001, *A&A*, **374**, 656.
Glenn, J., et al. 1994, *ApJ*, **424**, 967.
Goldreich, P & Wu, Y. 1999, *ApJ*, **511**, 904.
Horne, K., Marsh, T.R., Cheng, F.R., Hubeny, I. & Lanz, T. 1994, *ApJ*, **426**, 294.
Howell, S.B., Gaensicke, B.T., Szkody, P. & Sion, E.M. 2002a, *ApJL*, **575**, 419.
Howell, S.B., Harrison, T.E. & Osborne, H. 2002b, AAS Abstract, **201**, 4008.
Kawka, A. & Vennes, S. 2003, *AJ*, **125**, 1444.
Kawka, A. & Vennes, S. 2005, these proceedings.
Kovetz, A., Prialnik, D. & Shara, M.M. 1988, *ApJ*, **325**, 828.
Mukai, K. & Charles, P.A. 1986, *MNRAS*, **222**, 1P.
Naylor, T. 2002, In Classical Nova Explosions, eds M. Hernanz & J. Jose, AIP Conf. Proc., **637**, 16.

O'Donoghue, D., Koen, C., Kilkenny, D., Stobie, R.S., Koester, D., Bessell, M.S., Hambly, N. & MacGillivray, H. 2003, *MNRAS*, **345**, 506.
Robinson, E.L., et al. 1995, *ApJ*, **443**, 295.
Schwope, A.D. & Beuermann, K. 1993, In White Dwarfs: Advances in Observation & Theory, ed. M.A. Barstow, Kluwer, 381.
Schwope, A.D., Hambaryan, V., Schwarz, R., Kanbach, G. & Gaensicke, B.T. 2002, *A&A*, **392**, 541.
Shara, M.M. 1989, PASP, **101**, 5
Shara, M.M., Prialnik, D. & Kovetz, A. 1993, *ApJ*, **406**, 220.
Sion, E.M. 1991, *AJ*, **102**, 295.
Sion, E.M. 2003, in *NATO ASIB Proc. 105: White Dwarfs*, 303.
Sion, E.M., et al. 1995a, *ApJ*, **439**, 957.
Sion, E.M., Szkody, P., Cheng, F. & Min, H. 1995b, *ApJL*, **444**, 97.
Sion, E.M., Szkody, P., Cheng, F., Gaensicke, B.T. & Howell, S.B. 2003, *ApJ*, **583**, 907.
Stephenson, R.F. 1986, In RS Ophuichi and the Recurrent Nova Phenomenon, ed. M.F. Bode, VNU Sci. Press (Utrecht), 105.
Szkody, P., Bailey, J.A. & Hough, J.H. 1983, *MNRAS*, **203**, 749.
Szkody, P., et al. 2002a, *AJ*, **123**, 430.
Szkody, P., Gaensicke, B.T., Howell, S.B. & Sion, E.M. 2002b, *ApJ*, **575**, 79.
Szkody, P., Sion. E.M., Gaensicke, B.T. & Howell, S.B. 2002c, ASP Conf. Ser., **261**, 21.
Szkody, P., et al. 2003, *AJ*, **126**, 1499.
Townsley, D.M. & Bildsten, L. 2003, *ApJL*, **596**, 227.
Verbunt, F. 1984, *MNRAS*, **209**, 227.
Warner B., 2002, In Classical Nova Explosions, eds M. Hernanz & J. Jose, AIP Conf. Proc., **637**, 3.
Warner, B. & van Zyl, L. 1998, IAU Symp. No. 185, 321.
Warner, B. & Woudt, P.A. 2003, Proc. IAU Colloq. 193, in press.
Warner, B., Livio, M. & Tout, C.A. 1992, *MNRAS*, **282**, 735.
Wickramasinghe, D.T., Visvanathan, N. & Tuohy, I.R. 1984, *ApJ*, **286**, 328.
Wood, J.H., Horne, K. & Vennes, S. 1992, *ApJ*, **385**, 294.
Woudt, P.A. & Warner, B. 2004, *MNRAS*, **348**, 599.

WHITE DWARFS IN CATACLYSMIC VARIABLES: PROBES OF ACCRETION HISTORY

Review paper

Paula Szkody,[1] Edward M. Sion,[2] and Boris T. Gänsicke[3]

[1]*Department of Astronomy*
University of Washington, Box 351580
Seattle, WA 98195
szkody@astro.washington.edu

[2]*Department of Astronomy and Astrophysics*
Villanova University, 800 Lancaster Ave
Villanova, PA 19085
edward.sion@villanova.edu

[3]*Department of Physics*
University of Warwick
Coventry CV4 7AL, UK
Boris.Gaensicke@warwick.ac.uk

Abstract The white dwarfs in accreting close binaries provide a direct probe into the effects of magnetic fields and long term accretion on the evolution of these systems. Spectra from HST, FUSE, and the Sloan Digital Sky Survey reveal the range of temperatures, the rotation rates and the accretion geometries that are linked to the magnetic fields and the ages of the white dwarfs. Identifying these parameters allows us to obtain clues about how the evolution of single white dwarfs differs from those in close binary environments and how the cooling sequence is affected by the mass transfer.

Keywords: cataclysmic variables,accretion,ultraviolet,spectra

1. Introduction

The evolutionary history of accreting close binaries usually involves an episode of a common envelope (formed when the more massive star fills its Roche lobe during its giant phase) resulting in the spiral-in of the binary (which brings the two stars close together), followed by mass transfer from the late type main sequence star to what has become a white dwarf once sufficient

E.M. Sion, S. Vennes and H.L. Shipman (eds.), White Dwarfs: Cosmological and Galactic Probes, 205-210.
© 2005 *Springer*.

angular momentum loss has occurred via magnetic braking or gravitaional radiation (e.g. Howell, Nelson & Rappaport 2001). This entire scenario means the white dwarfs in binaries should have different temperatures (heating by accretion), rotation rates (spin-up by accretion, spin-down by losses during novae explosions), compositions (from the common envelope and mass transfer) and possibly magnetic field geometries than their single counterparts. By determining the values of these parameters for the white dwarfs in accreting binaries, we can obtain some idea of what they have gone through in their past. This paper will focus on some recent results on temperatures, rotation rates and field geometry, while the paper by Sion in this volume will present the results on composition differences.

At the current time, the best way to determine the parameters of hot white dwarfs in cataclysmic variables (CVs) is to obtain spectra in the ultraviolet, where the white dwarf flux can dominate over an accretion disk or magnetic accretion column. The highest probability of viewing the white dwarfs exists for novalike systems during low states when the mass transfer has basically stopped, or particular dwarf novae which have very low mass transfer rates at quiescence (usually the shortest orbital period systems). The UV spectra with STIS on HST provide coverage from about 1150-1700Å at about 1Å resolution and FUSE provides spectra from 900-1180Å at resolutions up to 0.03Å. With these instruments, the UV spectra can be fit with white dwarf models, extracting some measure of the temperature, gravity, mass, rotation, magnetic field and distance, depending on the S/N. The coolest white dwarfs can be fit from the optical, again if the mass transfer rates are low enough so that the accretion disk/column are not dominating the light. The Sloan Digital Sky Survey (SDSS) is ideal for probing the faintest, low mass transfer rate CVs for visible white dwarfs.

Since month-year timescales involve large changes in the mass-transfer rates of CVs, i.e. during outbursts of dwarf novae and high/low state behavior of novalikes, obtaining spectra during these different states enables a study of how much the white dwarf is affected by the changing accretion. In particular, it is possible to determine how the long-term evolutionary cooling sequence of a white dwarf is affected by short term accretion heating. By comparing accretion heating from different geometries (via an accretion disk which creates a boundary layer in a ring around the white dwarf or via an accreting magnetic pole which creates a small heated spot on the white dwarf), the effects of the magnetic field on the long term evolution can also be determined.

2. Information from Low States and Quiescence

The dominance of the white dwarf in the UV spectra of some dwarf novae at quiescence first became evident with IUE spectra of U Gem and VW Hyi (Panek & Holm 1984; Mateo & Szkody 1984). However, it took the greater sensitivity and resolution of HST to explore the white dwarfs in a variety of systems, especially those with short orbital periods (Sion 1999; Szkody et al. 2002). These spectra have shown that some white dwarfs such as LL And and EG Cnc can account for all of the observed UV flux near Lyα, while others, such as SW UMa and WX Cet need some disk contribution to provide the best match to the spectrum. The coolest white dwarf found in all of the STIS spectra exists in EG Cnc, with a temperature of 12,300K. The hottest white dwarf from STIS (46,000-50,000K) was revealed during a serendipitous observation that occurred during a low mass transfer state of the novalike system DW UMa (Knigge et al. 2000; Araujo-Betancor et al. 2003). In this case also, the white dwarf contributes all the UV flux. FUSE has also revealed hot white dwarfs in long period dwarf novae systems such as RU Peg and SS Aur (Sion et al. 2004), although the disk also contributes a substantial amount in these systems. A bare, hot (47,000K) white dwarf is apparent in the FUSE spectrum of the novalike MV Lyr during a low state (Hoard et al. 2004). In this system, the combination of the FUSE and optical spectra allows an upper limit to the mass transfer rate of $<3\times10^{-13}M_\odot$/yr to be determined.

Fewer magnetic systems have good STIS or FUSE spectra available and the results are harder to interpret. The system with the highest magnetic field of 240 MG (AR UMa) shows Zeeman split components of Lyα, as well as a forbidden transition of H (Gänsicke et al. 2001). A rough temperature of about 20,000K appears consistent with the UV spectrum but the entire UV-optical continuum does not fit well with any magnetic model. The FUSE spectrum also shows forbidden transitions of Lyβ near 1050 and 1100Å and the shapes of these features imply a displaced dipole. The coolest white dwarf in any CV is evident in a magnetic system found in the SDSS (SDSSJ1553+55; Szkody et al. 2003a). The very low temperature of 5000K for this system is consistent with very little accretion ($10^{-14}M_\odot$/yr) onto a white dwarf of over 4 Gyr age.

The resulting best-constrained temperatures obtained during quiescence and /or low states for accretion disk and magnetic systems with orbital periods under 5.5 hrs are shown in Figure 2. A summary of the temperatures for the STIS programs of Cycles 7 and 8 is also given in Table 1. Four interesting points emerge from inspection of Figure 2:

- 1. Temperatures of accreting white dwarfs are greater than single WDs
- 2. WD temperature increases with orbital period (accretion rate)
- 3. The lowest temperature WDs are all in magnetic systems

- 4. Temperatures are abnormally high in the 3-4 hr period range

Figures such as this confirm that long-term accretion does heat the white dwarf to values above what is expected from the normal cooling sequence. This effect is most significant for higher accretion rates, especially in the 3-4 hr period range where the high accretion rate systems termed SW Sex stars exist. It is also apparent that accretion at a magnetic pole is not as efficient at long-term heating as is accretion via a disk.

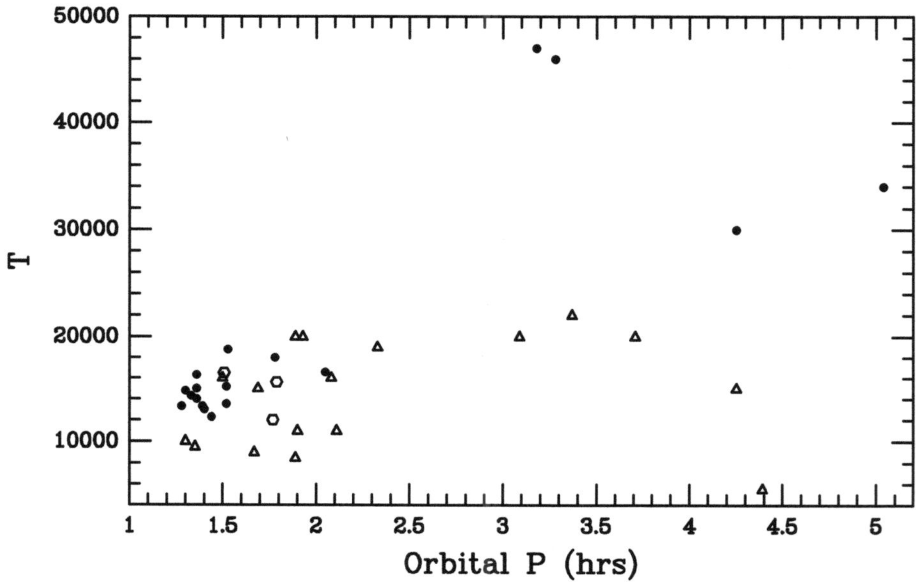

Figure 1. WD temperatures vs Orbital Period. Circles are systems with accretion disks (filled are HST, FUSE data; open are IUE), triangles are systems with highly magnetic WDs.

Table 1 also lists the rotation rates determined primarily by comparing the observed SiII and CI absorption lines in the spectra of the systems with adequate S/N to models that are broadened with rotation velocities ranging from 200-800 km/s. The results so far show that the rotation velocites are certainly higher than single white dwarfs (30 km/s) but appear to be much less than breakup velocity (most systems are not eclipsing but average inclinations are assumed). Thus, it is apparent that the spin-up from accretion is counterbalanced by spin losses through other mechanisms.

3. The Effect of Outbursts

The best opportunity to study the effects of outbursts on accretion heating of the underlying white dwarf occurs in the systems with large amplitude,

Table 1. STIS Results.

Object	P(hrs)	T_{wd}	d(pc)	Vsini(km/s)	Reference
CP Eri	0.48	(22000)			
GW Lib	1.28	13300	150	<300	ApJ 575, 79L 2002
BW Scl	1.30	14800	131		
LL And	1.33	14300	767	<500	ApJ 575, 419 2002
AL Com	1.36	16300	845	<800	AJ 126, 1451 2003
SW UMa	1.36	14000	159	200±50	
HV Vir	1.39	13300	510	400±100	ApJ 574, 950 2002
WX Cet	1.40	13000	187	400±100	ApJ 583, 907 2003
EG Cnc	1.44	12300	421	650±150	ApJ 574, 950 2002
VY Aqr	1.52	13500	133	400±100	ApJ 583, 907 2003
BC UMa	1.52	15200	287	300±100	
EK TrA	1.53	18800	210	200±100	AAp 374, 656 2001
AR UMa	1.93	20000	88		ApJ 555, 380 2001
EF Peg	2.05	16600	399	300±100	ApJ 575, 419 2002
DW UMa	3.28	46000	830	370±100	ApJ 539, L49 2000

long lasting outbursts. These systems are referred to as WZ Sge or TOADs (Warner 1995 reviews all characteristics). Although these outbursts happen infrequently, the prototype, WZ Sge, underwent one in July 2001 which was well-studied by HST and FUSE, as well as with ground-based and X-ray satellites. The FUSE (Long et al. 2003) and HST (Sion et al. 2003; Long et al. 2004) coverage over many months showed the initial disk and wind-dominated UV spectra changing over to a white-dwarf dominated spectrum as the disk returned to its quiescent configuration. While temperature fits to the white dwarf are complicated by intervening material close to outburst, it is clear that the white dwarf is definitely heated by the outburst and subsequently cools over long timescales. The HST data show temperatures near 23000K at 2 months after outburst which decrease to 16000K at 19 months past outburst. Even at 19 months, the UV flux was still 2.4 times higher than prior to outburst, so equilibrium had not yet been reached. Long term HST spectra following the outburst of another WZ Sge-type system, AL Com, also showed that the UV flux continued to decline for more than a year past outburst (Szkody et al. 2003b). However, in this case, the UV flux declined by a factor of 2 between 1 and 5.5 yrs past outburst, but the flux distribution remained identical, implying either a change in size (not temperature) of the emitting region or multiple components that happened to change in ways that resulted in similar distributions. Although the details of all the components contributing to the UV flux in outburst are not clear, it is very obvious that significant accretion heating does occur during outbursts and that it takes years to reach an equilibrium quiescent temperature.

An interesting caveat to the outburst luminosity (powered by accretion) is the work on HST involving parallaxes which determine distance (Harrison et al. 2004). These distances have provided a better M_V-P_{orb} relation and shown that the outbursts of WZ Sge type systems are 1.5-3 mag more luminous than normal outbursts. Their increased and longer-lasting heating during individual outbursts is counter-balanced by the low frequency of their outbursts (tens of years). If we assume a 2 mag increase in brightness (factor of 6 in accretion luminosity) and an outburst that lasts 10 times longer than for a typical dwarf nova, then there should be equivalent heating of a WZ Sge white dwarf as for a dwarf nova that erupts about 3 times a year (e.g. U Gem). But the white dwarf in U Gem has a temperature near 30,000K while WZ Sge is near 15,000K at quiescence. Hence, there is either some additional heating from the larger accretion rate of U Gem between outbursts, or the white dwarf in U Gem is younger than that in WZ Sge.

Acknowledgments

This work was supported by NASA through grants GO-0813.01-97A from the Space Telescope Science Institute, which is operated by the Association of Universities for Research in Astronomy, Inc., under NASA contract NAS5-26555, by NASA FUSE grant NAG5-10 and NSF grant AST-02-05875 to the University of Washington and NASA FUSE grant , and NSF grant to Villanova. B.T.G. acknowledges support from a PPARC Advanced Fellowship.

References

Harrison, T. E. et al. 2004, *AJ*, **127**, 460.

Hoard, D. W., Linnell, A., Szkody, P., Fried, R. E., Sion, E. M., Hubeny, I. and Wolfe, M. 2004, *ApJ*, **604**, 346.

Howell, S. B., Nelson, L. A. and Rappaport, S. 2001, *ApJ*, **550**, 897.

Long, K. S., Froning, C. S., Gänsicke, B. T., Knigge, C., Sion, E.M. and Szkody, P. 2003, *ApJ*, **591**, 1172.

Long, K. S., Sion, E. M., Gänsicke, B. T. and Szkody, P. 2004, *ApJ*, **602**, 948.

Mateo, M. and Szkody, P. 1984, *AJ*, **89**, 863.

Panek, R. J. and Holm, A.V., 1984, *ApJ*, **277**, 700.

Sion, E. M. 1999, *PASP*, **111**, 532.

Sion, E. M. et al. 2003, *ApJ*, **592**, 1137.

Sion, E. M., Cheng, F.H., Godon, P., Urban, J. and Szkody, P. 2004, *AJ*, **128**, 1834.

Szkody, P., Sion, E. M., Gänsicke, B. T. and Howell, S. B. 2002, *ASP Conf. Ser.*, **261**, 21.

Szkody, P., et al. 2003a, *ApJ*, **583**, 902.

Szkody, P., Gänsicke, B. T., Sion, E. M., Howell, S. B. and Cheng, F.,H. 2003b, *AJ*, **126**, 1451.

INTERMEDIATE POLARS IN LOW STATES

Brian Warner

Department of Astronomy, University of Cape Town, Rondebosch 7700, South Africa
warner@physci.uct.ac.za

Abstract Although no intermediate polar (IP) has been observed in recent years to descend into a state of low rate of mass transfer, there are candidate stars that appear to be already in intermediate and low states. V709 Cas is probably an intermediate state IP; NSV 2872 appears to be a low state IP, and will probably resemble the long orbital period IP V1072 Tau if it returns to a high state. The enigmatic star V407 Vul, which has had many interpretations as an ultra-short period binary, has many resemblances to the pre-cataclysmic variable V471 Tau, and may therefore be an IP precursor of quite long orbital period, and not yet a fully fledged cataclysmic variable.

Keywords: cataclysmic variables, close binaries, photometry

1. Introduction

The two principal classes of magnetic cataclysmic variable (mCV) are the synchronous rotators and the asynchronous rotators. The former, known as polars, possess white dwarf primaries with fields strong enough to interact with the secondary star and effect synchronous rotation of the primary with the orbital motion of the secondary. In the latter, known as intermediate polars (IPs) the primary is less strongly magnetized so the primary rotates with a period different from P_{orb}.

The polars are well known for having unstable mass transfer from the secondary to the primary – with resulting high and low states of accretion luminosity. Although relatively low states of IPs are known from studies of archived plates (Garnavich & Szkody 1988), and V1223 Sgr in particular has had extensive low states, it is a curious, and often regretted, fact that since they were recognized in the early 1980s, none of the well-studied IPs has entered into a low state. There would be obvious advantages to studying IPs in low states, particularly if the rate of mass transfer $\dot{M}$ were to shut off completely – the spectrum of the white dwarf would then be visible uncontaminated by accre-

E.M. Sion, S. Vennes and H.L. Shipman (eds.), White Dwarfs: Cosmological and Galactic Probes, 211-215.
© 2005 *Springer*.

tion emission and the possibility of measuring field strengths via the Zeeman effect, as is done during polar low states, would appear.

2. V709 Cas

A first move in this direction has been made possible by V709 Cas, which has P_{orb} = 5.34 h and P_{rot} = 5.22 min. The optical spectrum of this system shows the presence of broad absorption lines of the white dwarf primary, on which are superimposed the continuum and emission lines from the accretion disc and accretion curtains (Bonnet-Bidaud et al. 2001). Only an upper limit $B < 10$ MG is so far possible for the field strength, but variable X-Ray flux on time scales of months to years show that V709 Cas has unstable $\dot{M}$ which may at some time fall low enough for the white dwarf spectrum to be more clearly studied. Both Bonnet- Bidaud (2001) and De Martino et al. (2001), from different approaches, find $\dot{M} \sim 1 \times 10^{16}$ g s^{-1}, which is a factor of 4 – 10 lower than the values for $\dot{M}$ deduced for more normal IPs (see Table 7.4 of Warner 1995).

Thus V709 Cas appears to be an IP in an intermediate state of Mdot, which is compatible with the weakly visible spectrum of the primary. Its present brightness is $m \sim 14$, so a reconstruction of its historic light curve from archived plates is both feasible and desirable – if it attains a normal high state it could reach 12th magnitude.

3. V1072 Tau and NSV 2872

As a second step we might wonder whether there are in fact IPs in long-lived low states, already known but not properly recognized as such. As an example we can consider the IP V1072 Tau and ask what it would look like if it were to descend to a state of low $\dot{M}$.

V1072 Tau is a long period IP, with P_{orb} = 9.95 h and P_{rot} = 62.0 min (Remillard et al. 1994). Its optical spectrum is that of a late K star with superimposed emission lines and continuum from the accretion process. The K spectral type of the secondary is appropriate for a Roche lobe filling star in long period orbit. If the system were in a low state the spectrum would be dominated by the K star absorption spectrum, but there would be a contribution at short wavelengths (the U band, and shorter wavelengths), with the flux modulated at 62 min period either from the low level of Roche lobe overflow, or from magnetically channeled wind from the secondary (as in V471 Tau, which we discuss below).

Such an object might be quite difficult to discover (but would certainly be found eventually in any wide field survey like the Sloan Digital Sky Survey). It could, however, be readily found if the long-term light curve shows higher states of $\dot{M}$.

The object described here sounds very like the recently recognized low state CV known as NSV 2872. This star was first found to be variable by Ruegemer (1933), and a more complete light curve was provided by Zinner (1932) and Florja & Kukarkin (1935). It has a brightness range of m_{pg} = 11.4 – 14.5 and was classified in the GCVS as a suspected nova-like or dwarf nova. A spectrum obtained near the lower end of its brightness range showed only the absorption lines of an early K star, with no emission lines, and was consequently thought not to be a CV (Liu & Hu 2000). However, Kozhevnikov (2003) has made extensive photometric observations near minimum brightness (at $B \sim 14.4$: Kozhevnikov, private communication) and concludes that it (a) has very low level rapid flickering, characteristic of a CV in a low state of $\dot{M}$ and (b) has a coherent periodicity at 87.850 min of low and variable amplitude (3 – 8 mmag).

NSV 2872 therefore looks very much what we would expect of an IP in a long orbital period, accreting at a very low rate. The already recognized range of brightness of $\sim$ 3 mag shows that it can have high states. This is another system for which a more complete historical light curve would be very useful – and one that should be monitored regularly in order to detect the occurrence of a high state, where it could take on the appearance of V1072 Tau. A far UV study of the hot component is also obviously of some interest.

4. V471 Tau

The eclipsing binary V471 Tau is a member of the Hyades cluster and is a detached pair consisting of a 35 000 K white dwarf in orbit around a dwarf K2 secondary with P_{orb} = 12.5 h. What makes it more interesting is that it shows a 9.25 min modulation in the U band with an amplitude of 9.5 mmag (Clemens et al. 1992) that is 180° out of phase with pulsed soft X-Rays at the same period. From the known contribution of the white dwarf to the U band we can estimate that the true amplitude of the 9.25 min modulation is $\sim$ 30 mmag.

The explanation of the modulation is that the white dwarf has a magnetic field sufficiently strong to capture and channel part of the stellar wind from the secondary (Clemens et al. 1992). Thus V471 Tau is really a pre-IP, which will become an IP as soon as the secondary begins to overflow its Roche lobe.

The optical spectrum of V471 Tau shows the continuum and absorption lines of a K2 dwarf, currently with no emission lines (though in past years there has been Hα in emission, around phase 0.5 of the orbit) – Rottler et al. (1998). In the UV, of course, a continuum contribution from the white dwarf is present.

There are strong similarities between V471 Tau and NSV 2872 – the difference being that the former does not have the low level flickering that is probably the signature of Roche lobe overflow.

5. V407 Vul

V407 Vul has had a roller coaster career. Discovered initially as an X-Ray source, RX J1919.4+2456, modulated at 569 s period and thought to belong to the class of soft X-Ray IPs (Haberl & Motch 1995; Motch et al. 1996), it was later suggested that it could be a strongly magnetic He-transferring system (i.e., an AM CVn polar) where the observed period would be orbital (Cropper et al. 1998). The optical component was later identified, at $I \sim 18.6$, and was found to be modulated at the 9.5 min period, being in antiphase with the X-Ray variation (Ramsay et al. 2000). The X-Ray flux was found to vary by an order of magnitude on a time scale of a year. Later observations (Ramsay et al. 2002) showed modulation in the R and V bands, the latter with an amplitude of ~ 65 mmag, and a spectral flux distribution like that of a reddened K star, with no emission lines.

Alternate models were proposed for V407 Vul, to account for the lack of emission lines and absence of polarization. Marsh & Steeghs (2002) suggested that the inter-star stream of gas impacts directly onto the primary, as in an Algol system, thus preventing an accretion disc from being formed. A unipolar-inductor model, as in the Jupiter-Io system, was proposed by Wu et al. (2002).

With the acquisition of a higher quality spectrum of V407 Vul it is now known that the spectrum is simply that of an apparently normal reddened K star, with the addition of a featureless pulsed component at the shortest wavelengths (Steeghs 2003). There are no emission lines and there is no helium present.

The similarities of the spectrum, X-Ray and pulsation period of V407 Vul and V471 Tau are quite striking. There are parallels, too, with the properties of NSV 2872.

6. Conclusions

Although no recognized IP has entered a low state since this class of mCV was identified, there appear to be low $\dot{M}$ IPs available for study. V709 Cas has the signature of an IP in an intermediate state, and will repay study if it moves up or down from that state. NSV 2872 has features that would be expected of a long orbital period IP in a low state – in particular, if it returns to a high state it should resemble the high $\dot{M}$ IP V1072 Tau.

The enigmatic object V407 Vul, which has had many interpretations, has a strong resemblance to V471 Tau and so strictly is probably not a CV at all – it may be a long period pre-CV which will become an IP. Such systems may be relatively common – they are not easy to detect optically because they are dominated by the K star flux, yet the brightest is as close as the Hyades cluster. Lower mass systems, with M spectral type companions, will be easier to find as White Dwarf/M dwarf pairs, but will require appropriate extended photometry to detect rotationally modulated flux. Similarly, new detections through discovery of periodically modulated X-Ray flux will require extended pointed observations.

Acknowledgments

I thank Dayal Wickramasinghe for helpful discussions. My research is funded by the University of Cape Town.

References

Bonnet-Bidaud, J.M., Mouchet, M., de Martino, D., Matt, G. & Motch, C. 2001, *A&A*, **374**, 1003.
Clemens, J.C., et al. 1992, *ApJ*, **391**, 773.
Cropper, M., et al. 1998, *MNRAS*, **293**, 57L.
de Martino, D., et al. 2001, *A&A*, **377**, 499.
Florja, N.F. & Kukarkin, B.V. 1935, *Perem. Zvezdy*, **5**, 19.
Garnavich, P. & Szkody, P. 1988, *PASP*, **100**, 1522.
Haberl, F. & Motch, C. 1995, *A&A*, **397**, L37.
Kozhevnikov, V.P. 2003, *A&A*, **398**, 267.
Liu, W. & Hu , J.Y. 2000, *ApJS*, **128**, 387.
Marsh, T.R. & Steeghs, D. 2002, *MNRAS*, **331**, 7L.
Motch, C., et al. 1996, *A&A*, **307**, 459.
Ramsay, G., Cropper, M., Wu, K., Mason. K.O. & Hakala, P. 2000, *MNRAS*, **311**, 75.
Ramsay, G., et al. 2002, *MNRAS*, **333**, 575.
Remillard, R.A., et al. 1994, *ApJ*, **428**, 785.
Rottler, L., Batalha, C., Young, A. & Vogt, S. 1998, *BAAS*, **192**, 6720.
Ruegemer, H. 1933, *Astronomische Nachrichten*, **248**, 410.
Steeghs, D. 2003, Workshop on Ultracompact Binaries, Santa Barbara, see `http://online.kitp.ucsb.edu/online/ultra_c03/steeghs/`.
Warner, B. 1995, Cataclysmic Variable Stars, Cambridge University Press
Wu, K., Cropper, M., Ramsay, G. & Sekiguchi, K. 2002, *MNRAS*, **331**, 221.
Zinner, E. 1932, *Astronomische Nachrichten*, **246**, 17.

CHEMICAL ABUNDANCES OF WHITE DWARFS IN CATACLYSMIC VARIABLES

Edward M. Sion
Department of Astronomy and Astrophysics,
Villanova University, Villanova, PA 19085, USA
edward.sion@villanova.edu

Paula Szkody
Department of Astronomy,
University of Washington, Seattle, WA 98195, USA
szkody@astro.washington.edu

Abstract

There is spectroscopic evidence suggesting the past occurences of thermonuclear runaways (TNRs) on the accreting white dwarfs in certain dwarf novae both above and below the orbital period gap. In the dwarf novae U Gem, VW Hyi and WZ Sge, synthetic spectral analyses of HST, FUSE and IUE archival spectra indicate supra-solar N and depleted C in their WD photospheres during dwarf nova quiescence. Sub-solar metal abundances ($\sim 0.1 to 0.5\times$ solar) have been reported in the white dwarf photospheres of 14 dwarf novae below the period gap but N abundances are lacking. Several CV white dwarfs reveal elevated abundances of odd-numbered nuclear species P, Mn and Al indicative of proton capture by even-numbered nuclear species outside the CNO by-cycle during the CNO H-burning thermonuclear processing. In addition, there are magnetic and non-magnetic systems having peculiar N V/C IV emission line ratios, associated with the accretion disk or accretion column. Possible nova shells have been detected around two dwarf novae reinforcing the evidence from chemical abundances that dwarf novae and classical novae are spectroscopically linked and can overlap.

Keywords: Accretion, Dwarf Novae, White Dwarfs, Chemical Abundances

1. Introduction

Due to their enormous surface gravities and hence very short gravitational diffusion timescales, single white dwarfs display spectra that are essentially mono-elemental with the lightest element in the envelope always appearing at

E.M. Sion, S. Vennes and H.L. Shipman (eds.), White Dwarfs: Cosmological and Galactic Probes, 217-223.
© 2005 *Springer.*

the surface. All heavier elements have diffused downward. For a typical DA white dwarf with log g = 8 and $T_{eff} = 15,000$ K (the highest T_{eff} H-rich case tabulated by Paquette et al. 1986), the diffusion timescale for metals should be shorter than 0.011 to 0.007 year (i.e., $< 3-4$ days). The accreted atmospheres of CV white dwarfs are affected by thermal and gravitational diffusion, turbulent mixing and (for objects above 20,000K) possibly also by radiative forces. Hence, any metals detected in their photosphere cannot have remained there more than a few days. For white dwarfs in cataclysmic variables, the presence of heavy elements in the atmosphere is continually being replenished by disk accretion and presumably dredge-up from deeper layers where nucleosynthetic products have settled. A likely candidate mechanism for the dredge-up may be the dwarf nova explosion itself and its associated tangential accretion with shear mixing which could stir up deeper layers of the envelope.

The accretion of material from the secondary during the dwarf nova with no mixing would tend to cover over the TNR–processed material. The secondary is normally expected to consist of solar composition. It may, however, be contaminated by captured processed material during the nova ejection and brief common envelope stage following the TNR. Since it would be mixed with secondary solar material, the original nova would be even more enhanced than observed now.

2. Spectroscopic Clues from Dwarf Novae as Past Classical Novae

Dwarf novae and nova-like variables contain accreting white dwarfs which may have undergone numerous thermonuclear runaways as classical novae. Recently, two dwarf nova have been found to have nova shells which demonstrate their connection with classical novae. A connection between the two classes was already reported on the basis of chemical abundances by Sion et al. (1997). They showed that the surface abundances of the white dwarf in VW Hydri during its quiescence manifests a direct evolutionary link to a past thermonuclear event. This conclusion is based upon the presence of a large abundance ratio of nitrogen to carbon and the spectroscopic presence of odd-numbered proton capture nuclei in abundances greatly elevated above solar. Both of these spectroscopic characteristics point to hot CNO processing as the source of the abundances. In addition to VW Hyi, the dwarf nova U Gem has also been studied extensively with HST and manifests some of the same surface abundance characteristics as VW Hyi. We have selected VW Hyi as the best case example of securing abundances in an accreting white dwarf atmosphere.

In Figures 1 and 2 we display the HST STIS E140M spectra of VW Hyi obtained 2 days and 7 days following a superoutburst, compared with the best-fitting single temperature white dwarf. The resulting parameters and chemical abundances are summarized in Table 1 (see also Sion et al. 2001).

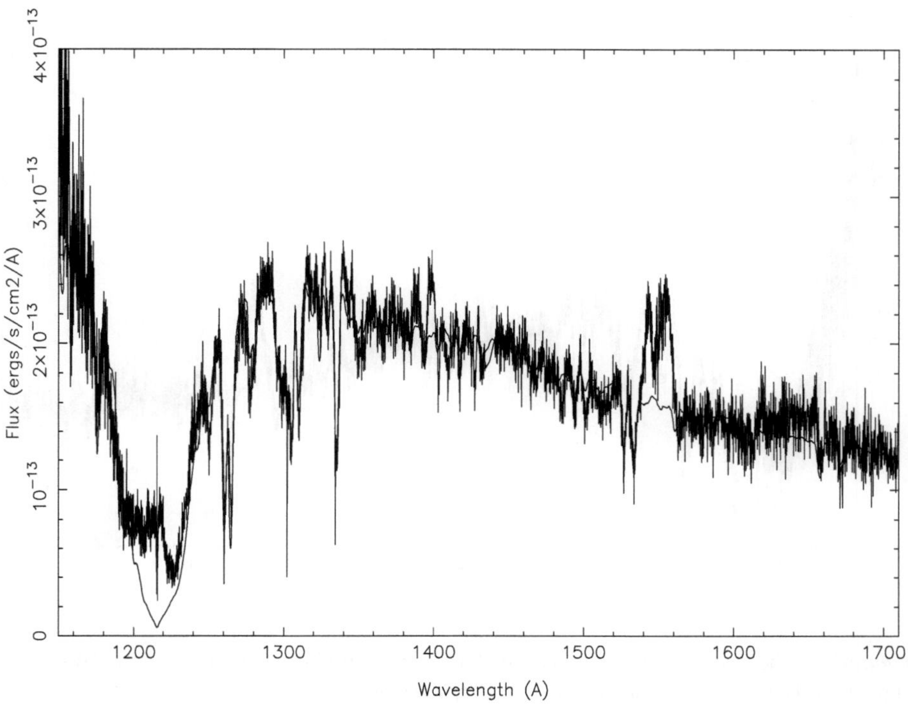

Figure 1. The best fit synthetic spectrum of a white dwarf alone, to spectrum 1, 2 days post-superoutburst ($T_{eff} = 22,500$K, log g = 8, $V_{sin} = 400$ km/s).

Table 1. Metal Abundance Differences

Obs. #	*V*	*Si*	*C*	*N*	*O*	*Al*	*Fe*
1	400	0.3	0.3	3	3	2	0.5
2	500	0.4	0.4	4	4	2	0.05
Obs #	*Mg*	*Mn*	*Ni*	*P*	*Ti*	χ^2	Scale
1	3	50	0.3	15	0.1	1.7110	3.9916×10^{-2}
2	5	50	0.3	20	0.4	4.9370	4.0693×10^{-1}

There are two metal abundance differences which could be statistically significant. The Fe and P reveal the largest difference in abundance between the two observations. The odd-numbered element phosphorus has an abundance

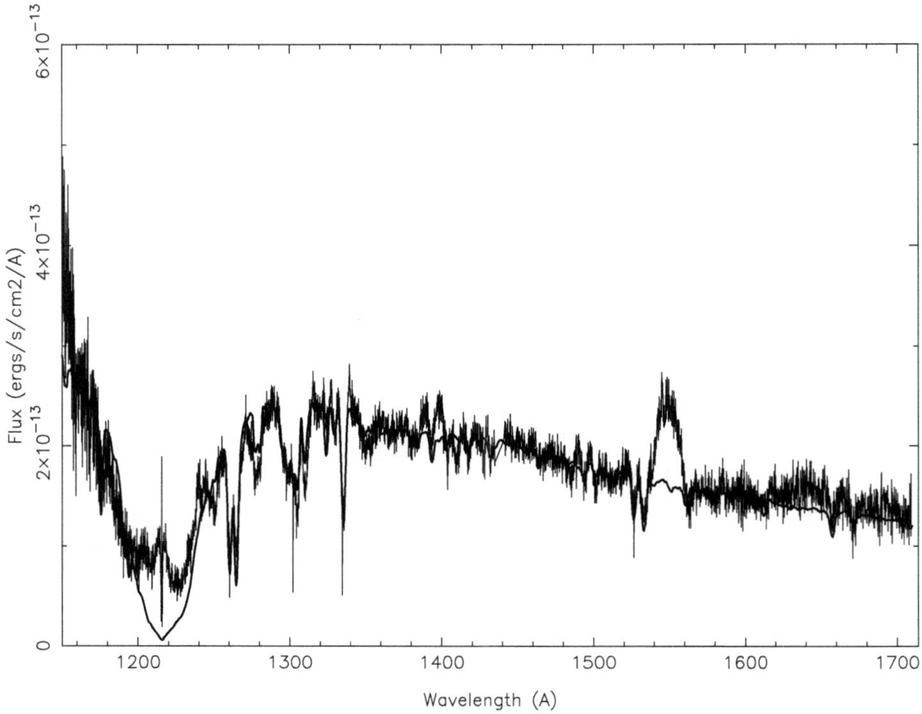

Figure 2. The best-fit synthetic spectrum of a white dwarf alone, to spectrum 2 of VW Hydri obtained 7 days post-superoutburst ($T_{eff} = 22,500$K, log $g = 8$, $V_{sin} = 500$ km/s).

of 15 times solar in spectrum 1 compared to 20 times solar in spectrum 2. The odd-numbered element Manganese is elevated to 50 times solar in both observations. However, the difference in the Fe abundance between the two observations is significantly larger. In obs. 2, the Fe abundance has declined by a factor of 10. We have no explanation for this other than diffusion. All of the other metal species show slight increases in spectrum 2 but the differences are probably too small to be significant. Moreover, any difference in abundances must be viewed cautiously since underlying disk emission during quiescence could be filling in the absorption making it difficult to secure reliable abundances for either spectrum.

We believe that the above two STIS observations strengthen the evidence that a TNR induced by the accretion of material from the secondary has occured in the past on the white dwarf of VW Hyi for the following reasons. (1) Confirmation of the elevated abundances of the odd-numbered nuclear species P, Al and Mn relative to their even-numbered nuclear neighbors. (2) The absolute abundance of nitrogen is larger than carbon. Solar composition contains 4 times as much carbon as nitrogen. During a TNR the carbon will capture protons and become nitrogen (Starrfield et al. 1978). Even a TNR on a ONeMg white dwarf will end up with more N than C as the O proton captures, ejects He, and becomes N (Politano et al. 1995). While the CNO nuclei are capturing protons during a TNR, the heavier nuclei will also be capturing protons. Because of the very short time that the temperature is high during a TNR, most nuclei can only capture a few protons. Thus, the even-numbered nuclei peaks in a solar abundance distribution will tend to be leveled out by filling in the valleys occupied by the odd-numbered nuclei (Na, Al, P, Mn). Since the even-numbered nuclei are 10 to 100 times more abundant than their odd-numbered neighbor nuclei, a smoothing of these peaks and valleys would lead to $\sim$ 0.5 solar abundances for the even- numbered nuclei and 5 to 50 times solar abundances for the odd-numbered nuclei. The actual value also depends on the proton capture rates and the number of stable states. This simple picture explains approximately the observed abundances in both of the STIS spectra. Hydrodynamic simulations with nucleosynthesis are required for a more accurate comparison.

The evidence, based upon chemical abundances, that bona fide dwarf novae are linked to past nova explosions is also supported by recent imaging of possible nova shells around two dwarf novae, EY Cygni (Sion et al. 2004) and Z Cam (Shara et al. 2004). In the case of EY Cygni, there is an anomalous ratio of N/C with very strong N V disk emission and very weak C IV enmission and substantial evidence for a near-Chandrasekhar mass white dwarf.

In U Gem, Sion et al.(1998) proposed that sufficient contamination of the secondary star by the brief common envelope phases of numerous classical nova explosions would result in the secondary transferring material back to the white dwarf that was enriched by the products of CNO processing. They found the white dwarf in U Gem to have a mass $M_{wd} = 1.12 \; M_{\odot}$ and $K_1 = 107$ km/s. They predicted that the N abundance in the white dwarf photosphere would be several times solar. All these results, including the white dwarf mass, were confirmed in the HST G140L study of Long & Gilliland (1998).

Of greatest interest here are the abundances: C: 0.05 to 1 solar, N: 4.0 solar, Si: 0.4 to 1.3 solar, He: 1.0 solar and all other metals 1.0 solar. Thus, the white dwarf in U Gem presents another solid example of a large N/C ratio, well above the solar value. Since the secondary stars in these dwarf novae are of very low mass, it is unlikely that the large N/C ratio is associated with CNO processing intrinsic to the M dwarf components. (see below however).

Additional abundance information for 12 non-magnetic CV white dwarfs has been reported from the HST STIS Cycle 8 medium program of Szkody et al. (2002). For the white dwarfs in the dwarf novae EG Cnc, BW Scl, GW Lib, BC UMa, SW UMa, VY Aqr, WX Ceti, AL Com, LL And, HV Vir, EF Peg, and EK Tra, sub-solar abundances of C, Si and O ($\sim 0.3 \pm 0.1\times$ solar) have been determined while in several of these systems, the WD photospheres appear to have supra-solar abundances of Al. Unfortunately there is no reliable information on whether N is elevated relative to C in this sample of objects.

3. Spectroscopic Evidence from IUE Studies of Dwarf Novae

While the number of cataclysmics with exposed white dwarf photospheres (and hence the possibility of abundance analyses) is rather limited, the signature of CNO processing may also be found in the emission line spectra originating in the accretion disk or, in the case of magnetic CVs, the accretion column. The most prominent emission line in the far UV spectra of dwarf novae is virtually always C IV (1548, 1550). Remarkably, elevated N/C is also seen in emission (very strong N V, weak or absent C IV) in a number of CVs (both magnetic and non-magnetic; Gaensicke et al. 2003; Winter and Sion 2003; Dulude and Sion 2002). The question is, do these peculiar abundances originate in the WD or in the nuclear evolution of the secondary? In the latter case, Schencker et al. (2002) have argued that the anomalous N/C abundances may originate in a CV secondary originally more massive than the white dwarf ($M_2 > 1.5M_\odot$) which underwent unstable thermal timescale mass transfer as a supersoft X-ray binary, survived the rapid mass transfer episode, and has been peeled down to its CNO-processed core layers and is now transferring processed core material.

4. Conclusions

While it would seem that only a past thermonuclear event could be responsible for the peculiar abundances, there is some question as to whether the contamination of the secondary by the many brief common envelope transient stages of a nova would be sufficient to account for the observed abundances. It cannot be excluded, albeit unlikely, that thermonuclear processing occurs during the high accretion episodes of dwarf novae in outbursts and nova-like variables in high states. There have been no realistic simulations of this possibility.

For the nova-like variables, only a handful of systems have been observed in the low state of little or no accretion when the white dwarf can be directly observed. However, three objects have been analyzed and contain very hot white dwarfs; TT Ari, MV Lyra and DW UMa. The higher temperatures are presumably due to the higher time-averaged accretion rates. In all three cases, nitrogen abundances remain uncertain.

Overabundances and depletions of certain elements, and N/C ratios suggest processing by hot CNO burning occurs in many CVs. However, their accreted atmospheres are also affected by gravitational diffusion, turbulent mixing and (for objects above 20,000K) possibly also by radiative forces which can complicate the interpretation of the abundance pattern. There is solid evidence that the WDs in U Gem and VW Hyi both have N = 4-5 solar and C = 0.1 solar, and elevated N/C is also seen in emission (very strong N V, weak or absent C IV) in a number of CVs (both magnetic and non-magnetic; Gaensicke et al. 2003). The question is, do these peculiar abundances originate in the WD or in the secondary, which has been peeled away to its core by mass transfer? Is there a way to discriminate between the two possibilities? If these ratios and overabundances originated primordially in an originally more massive ($M_2 > 1.5M_\odot$) secondary, then that material is being transferred to the WD and has nothing to do with past novae. If they originated in novae which repeatedly polluted the secondary, then the secondary material is being transferred back to the WD. A CV in which N/C is elevated in the WD photosphere but the disk emission lines have normal strength (i.e. no anomalous N V/C IV) would support the nova contamination hypothesis.

Finally, several white dwarfs in SU UMa-type dwarf novae have elevated Al and P in the WD photosphere. With higher resolution, better signal-to-noise spectra, it will be feasible to look for specific abundance markers, such as elevated N/C, depleted C, over-abundant odd-numbered nuclear species like Al and P (products of proton captures during hot CNO burning).

References

Dulude, M., & Sion, E.M.2002, *BAAS*, **201**, 4014.

Gaensicke, B., Szkody, P., de Martino, D., Beuermann, K., Long, K., Sion, E.M., Knigge, C., Marsh, T., Hubeny, I. 2003, *ApJ*, **594**, 443.

Long, K., & Gilliland, R.1998, *ApJ*, **511**, 916.

Paquette, C., Pelletier, C., Fontaine, G., & Michaud, G. 1986, *ApJS*, **61**, 197.

Politano,M., Starrfield, S.G., Truran, J., Weiss, A., & Sparks, W.M.1995, *ApJ*, **448**, 807.

Schenker, K., King, A., Kolb, U., Wynn, G.A., Zhang, Z.2002, *MNRAS*, **337**, 1105.

Shara, M.M. et al. 2004, *BAAS*, **205**, #159.01.

Sion, E.M, Cheng, F., Sparks, W.M., Szkody, P., Huang, M., & Hubeny, I. 1997, *ApJL*, **480**, L17.

Sion, E. M., Cheng, F., Szkody, P., Gänsicke, B., Sparks, W., & Hubeny, I. 2001, *ApJ*, **54**, 127.

Sion et al. 1998, *ApJ*, **496**, 449.

Sion, E.M., Winter, L., Urban, J., Tovmassian, G., Zharikov, S., Orio, M.2004, *AJ*, **128**, 1795.

Starrfield, S.G., Truran, J., & Sparks, W.M. 1978, *ApJ*, **226**, 186.

Szkody, P., Sion, E.M., Gaensicke, B., Howell, S.B.2002, in The Physics of Cataclysmic Variables and Related Objects, eds. B. Gaensicke, K. Beuermann and K. Reinsch, ASP Conf.Ser., 261, 21

Warner, B. 1995, Cataclysmic Variable Stars, (Cambridge, GB: Cambridge University Press).

Winter, L., & Sion, E.M.2003, *ApJ*, **582**, 352.

AM CVn STARS IN THE UCT CCD CV SURVEY

Patrick A. Woudt and Brian Warner

Department of Astronomy, University of Cape Town, Rondebosch 7700, South Africa
pwoudt@circinus.ast.uct.ac.za, warner@physci.uct.ac.za

Abstract High speed photometry of the helium-transferring binary ES Cet – taken over a two-year period (2001 October – 2003 October) – shows a very stable photometric period of 620.211437 $\pm$ 0.000038 s, with a tentative indication of curvature in the O–C diagram suggesting a change in period at a rate of $\dot{P} \sim 1.6 \times 10^{-11}$. Phase-resolved spectroscopy of ES Cet obtained with the Hobby-Eberly Telescope shows a clear modulation on the photometric period, the assumed orbital period. We have followed a newly identified AM CVn star ('2003aw') photometricaly through its 2003 February/March outburst during which it varied in brightness over a range of V = 16.5 – 20.3; we find a superhump period of 2041.5 $\pm$ 0.3 s. Questions are raised about the reality of the detected spin-up in RX J0806 (Hakala et al. 2003; Strohmayer 2003).

Keywords: cataclysmic variables, close binaries, photometry, spectroscopy

1. Introduction

There are currently ten known unequivocal double degenerate interacting binaries (AM CVn stars), namely ES Cet, AM CVn, HP Lib, CR Boo, KL Dra, V803 Cen, CP Eri, '2003aw', GP Com and CE-315, ranging in orbital period (P_{orb}) from 10.3 – 65.1 min. These stars have proper spectroscopic and photometric credentials – their spectra show helium emission or absorption lines; no hydrogen can be present in these systems. There are two additional candidate AM CVn stars of suspected short orbital period, RX J0806 at P_{orb} = 5.35 min (Israel et al. 2002; Ramsay et al. 2002) and V407 Vul (Cropper et al. 1998) at P_{orb} = 9.49 min. Their classification as AM CVn stars is, however, not unambigious; there is some (tentative) evidence for the presence of hydrogen in the spectrum of RX J0806 (Israel et al. 2002), and the spectrum of V407 Vul is that of a K star (Steeghs 2003) making it appear like an intermediate polar precursor at quite long orbital period (Warner 2003). In this interpretation, the 9.49-min photometric and X-ray modulation is associated with the spin pe-

E.M. Sion, S. Vennes and H.L. Shipman (eds.), White Dwarfs: Cosmological and Galactic Probes, 225-232.
© 2005 *Springer*.

riod of the primary, not the orbital period. Table 1 lists all the AM CVn stars, including the two candidates.

Table 1. The AM CVn stars

Object	*V (mag)*	P_{orb} *(s)*	P_{sh} *(s)*	*References*
RX J0806	21.1	321.25^a		1, 2
V407 Vul	19.9	569.38^a		3
ES Cet	16.9	620.21144		4, these proceedings
AM CVn	14.1	1028.7	1051.2	5, 6
HP Lib	13.7	1102.7	1119.0	7, 8
CR Boo	13.0 – 18.0	1471.3	1487	9, 10
KL Dra	16.8 – 20	1500	1530	11
V803 Cen	13.2 – 17.4	1612.0	1618.3	12
CP Eri	16.5 – 19.7	1701.2	1715.9	13
'2003aw'	16.5 – 20.3		2041.5	14, these proceedings
GP Com	15.7 – 16.0	2974		15, 16
CE-315	17.6	3906		17, 18

[a]Not yet definitively established as orbital periods.
[1]Israel et al. (2002); [2]Ramsay et al. (2002); [3]Cropper et al. (1998); [4]Warner & Woudt (2002); [5]Solheim et al. (1998); [6] Skillman et al. (1999); [7]O'Donoghue et al. (1994); [8]Patterson et al. (2002); [9]Wood et al. (1987); [10]Patterson et al. (1997); [11]Wood et al. (2002); [12]Patterson et al. (2000); [13]Abbott et al. (1992); [14]Woudt & Warner (2003a); [15]Nather et al. (1981); [16]Marsh et al. (1991); [17]Ruiz et al. (2001); [18]Woudt & Warner (2002).

2. The UCT CCD CV Survey

The UCT CCD CV Survey is a high speed photometric survey of faint cataclysmic variable stars (CVs) using the University of Cape Town (UCT) CCD photometer (O'Donoghue 1995) in frame-transfer mode, in combination with the 1.0-m and 1.9-m reflectors at the Sutherland site of the South African Astronomical Observatory.

ES Cet. Initial high-speed photometry of ES Cet obtained during four nights in 2001 October (Warner & Woudt 2002) showed a clear modulation at 620.26 s – in the Fourier transform only the fundamental and its first three harmonics of the 620.26-s modulation were present. The spectrum of ES Cet (see Fig. 2) is dominated by He II emission lines, and hence its position amongst the AM CVn stars is secure. From the low mass ratio (q = 0.094), and the predicted rate of mass transfer $\dot{M}$ of $\sim 1 \times 10^{-8}$ $M_\odot$ y^{-1} at P_{orb} = 620 s (Warner 1995), one expects that the photometric modulation originates from superhumps due to tidal distortions in the accretion disc (Patterson et al. 2002). In this case, the orbital period would be a few per cent lower than the observed photometric modulation.

We have followed ES Cet photometrically over the last two years and the photometric modulation is surprisingly stable. If indeed it arises from a superhump modulation, it is the most stable superhump detected to date (Patterson, priv. comm.). The O–C diagram of all the UCT CCD photometry taken between 2001 October and 2003 October is shown in Fig. 1, phased according to the ephemeris given in Eq. 1. Even though there is some scatter in the O–C diagram, there are no substantial phase shifts or period changes (as might have been expected were the modulation due to superhumps).

$$\mathrm{HJD_{min}} = 245\,2203.3739512 + 0.0071783731(4)\,\mathrm{E} \qquad (1)$$

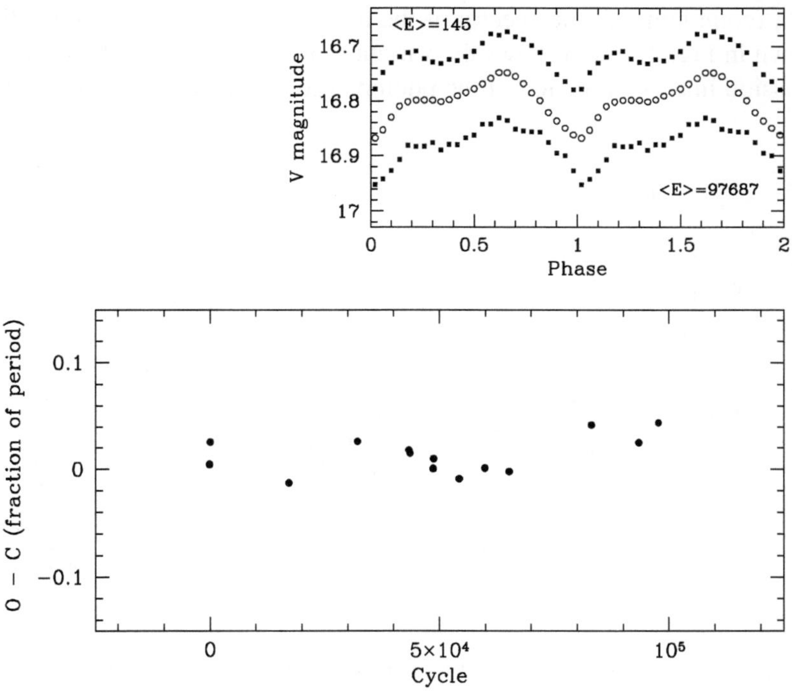

Figure 1. The O–C diagram (lower panel) of all the UCT CCD photometry obtained of ES Cet over the last 2 years. The upper panel shows the average light curve of all the data (open circles) compared with the mean light curve of two individual runs spaced two years apart.

With the two-year baseline, there is a slight hint of upwards curvature in the O–C diagram, which implies a lengthening of the photometric period. If the data are split in two halves (E $\sim$ 0 – 50 000 cycles and E $\sim$ 50 000 – 100 000 cycles, with an overlap of the dense data coverage at E $\sim$ 50 000 cycles), we find that the period is indeed larger in the second half: P_{orb} (1st year) = 620.211391 ($\pm$ 50) s versus P_{orb} (2nd year) = 620.211841 ($\pm$ 96) s. The amount of variation is consistent with the curvature seen in the lower panel of Fig. 1. It implies $\dot{P} \sim 1.6 \times 10^{-11}$ – fairly close to the expected rate of change of 6×10^{-12} (Warner & Woudt 2002) for a high $\dot{M}$ system. We realise that the data coverage is still rather small, and another two years of photometry will be required to confirm this trend in the O–C diagram.

Apart from the extended photometric coverage, we have obtained phase-resolved spectroscopy (with a time resolution of 30 s) of ES Cet using the Low Resolution Spectrograph on the Hobby-Eberly Telescope (HET) at the McDonald Observatory in Texas. Two visits of 40 minutes, and a third observing run 30 minutes long, showed very clearly that the spectral lines varied on the photometric period of 620.21 s. This has also been seen by Steeghs (2003) in two consecutive nights of phase-resolved Magellan data. The averaged HET spectra (combining all the spectra of the three different ES Cet observations) is shown in Fig. 2. Fig. 3 shows the variation of the centroid of the He II 4686 Å emission line as a function of the photometric ephemeris given in Eq. 1.

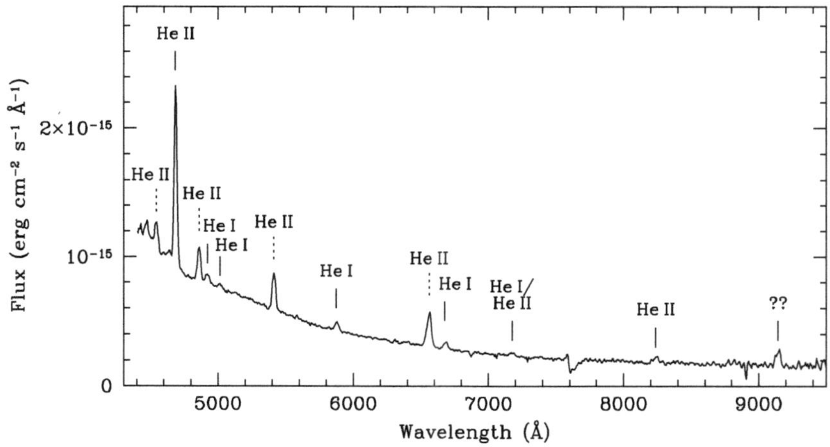

Figure 2. The averaged spectra of ES Cet taken with the Hobby-Eberly telescope.

The spectrum consists mainly of He I and He II emission lines (lines of the He II Pickering series are marked by the dashed vertical bars in Fig. 2); there is one line at $\lambda \sim 9140$ Å which we haven't yet been able to identify. The strong emission lines are somewhat unexpected for an object of inferred high $\dot{M}$. The low spectral resolution of the HET spectra fails to show the double-lined nature of the emission lines, but higher resolution spectra (Steeghs 2003) clearly show the double emission lines, indicating the presence of an accretion disc. The spectral resolution of the HET spectra is too low for generating Doppler tomograms, cf. Steeghs (2003).

Given the stability of the photometric period and spectroscopy modulation on the photometric period, it seems that the 620-s modulation is more probably the orbital period of the system and not the superhump period as commonly expected for a low-mass ratio, high $\dot{M}$ system.

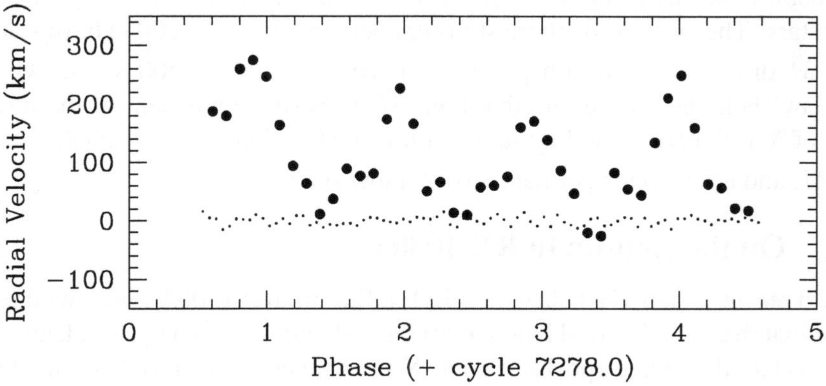

Figure 3. Variations in the centroid of the He II 4686 Å emission line (big dots), phased on the ephemeris given in Eq. 1. The small dots show the [O I] 5577 Å night sky line.

'2003aw'. Shortly after its outburst in 2003 February, '2003aw' was identified as a candidate supernova in a supernova search (Wood-Vasey et al. 2003), but a spectrum taken by Chornock & Filippenko (2003) revealed weak He I emission lines at zero redshift, making '2003aw' a candidate AM CVn star caught in a high state. Initial photometry with the UCT CCD confirmed this suggestion after finding a photometric modulation in the light curve with a period of 2034 $\pm$ 3 s (Woudt & Warner 2003b), with recurring dips in the light curve during the high state – possibly indicating shallow eclipses.

We followed '2003aw' through its decline from the high state into the intermediate state, and during both phases a photometric modulation of identical period was found: P_{sh} = 2041.5 $\pm$ 0.3 s (Woudt & Warner 2003a). The dips did not occur during the intermediate state. However, during the intermediate state we observed a 'cycling' in brightness of $\geq$ 0.4 mag on a time-scale of $\sim$ 16 h. This behaviour is also seen in other AM CVn stars in intermediate states; in CR Boo (P_{orb} = 1471 s) the cycle time is $\sim$ 19 h, with a range of 1.1 mag and in V803 Cen (P_{orb} = 1612 s) the cycle time is 22 h, with a range of 1.1 mag. '2003aw' has two other interesting aspects:

- The presence of sidebands to the fundamental *superhump* frequency and two of its harmonics during the high state. The frequency separation of these sidebands corresponds roughly to the 'cycling' time scale.

- A short lived ($\sim$ 1 day) brightening of ΔV $\sim$ 1.8 mag occured during the high state, resembling the behaviour of some intermediate polars (Schwarz et al. 1988; van Amerongen & van Paradijs 1989).

'2003aw' seems to know its place within the emerging hierarchy of AM CVn stars. The systems of shortest orbital periods ($P_{orb} \lesssim$ 1200 s) have stable high $\dot{M}$ discs, systems with periods between $\sim$ 1200 – 2500 s – to which '2003aw' belongs – have unstable high $\dot{M}$ discs (the equivalent of the nova-likes of VY Scl type), and systems with orbital periods $P_{orb} \gtrsim$ 2500 s have low $\dot{M}$, and are perhaps permanently in a low state.

3. On the spin-up in RX J0806

Two recent papers (Hakala et al. (2003); Strohmayer (2003)) presented evidence that the orbital period in RX J0806 is undergoing a spin up. Evidence for this was based on three epochs of data: X-ray data taken with ROSAT in 1994-1995 (Burwitz & Reinsch 2001), and two sets of optical data taken with the VLT and NOT (Hakala et al. 2003) in 2001 Nov/2002 Jan, and 2003 Jan/Feb, respectively. Of the two optical data sets, the first data set (2001/2002) suffers from severe aliasing (Hakala et al. 2003), and as a result, the deduced period evolution depends critically on the assumption that the X-ray period in the 1994/1995 ROSAT data set is correct.

The X-ray data (Burwitz & Reinsch 2001) were taken in 1994 October and 1995 April with a total of 13 400 s of integration time. This amounts to the equivalent of 42 cycles of the 5-min modulation spread out over $\sim$ 180 days ($\sim$ 50 000 cycles). With such poor data coverage, it is impossible to determine periods to the accuracy of 0.4 ms as quoted in Burwitz & Reinsch (2001) and perpetuated in Hakala et al. (2003). The aliasing is severe, as shown in Figure 4 of Burwitz & Reinsch (2001), and the highest peak in the forest of aliases is not necessarily the correct period. Each alias peak can be determined with a precision of 0.4 ms (largely determined by the baseline of the observations), but the choice of alias peak can lead to inaccurate results. To illustrate that, we have selected a few observing runs of ES Cet – mimicking an approximately similar data coverage (80 out of 32 000 cycles) – and show the FT for this small data set next to the complete data set in Fig. 4.

If the nearest alias to the right (see Figure 4 of Burwitz & Reinsch) of the preferred period in Burwitz & Reinsch (2001) is chosen (i.e., a lower period), the entire period evolution disappears (Figure 2 of Hakala et al. 2003). At best, the evidence for a spin-up in RX J0806 is tentative; there are currently insufficient data to claim (with high significance) a period evolution in this object. It remains to be seen if the proposed period evolution in V407 Vul (Strohmayer 2002) suffers from the same problem of poor data coverage (in

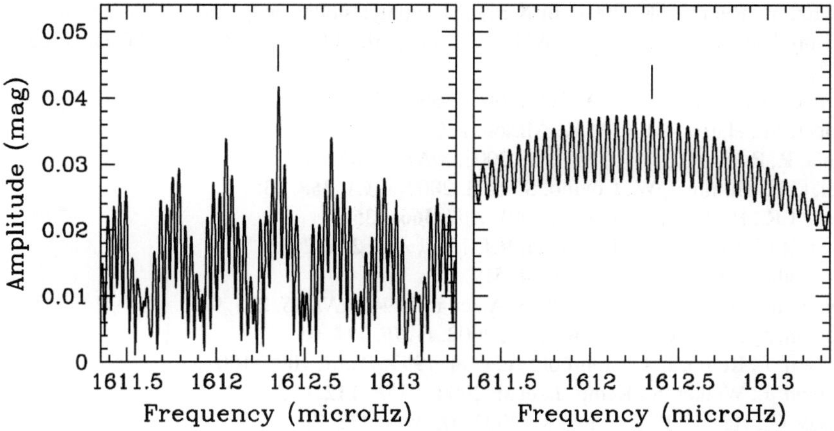

Figure 4. A comparison of the Fourier transform of all the SAAO observations of ES Cet in the period 2001 October – 2003 October (left panel) with a selected subsample with limited data coverage the equivalent of 80 cycles over 32 000 cycles (right panel). The proper photometric period is marked by the vertical bar in both diagrams.

that case there is X-ray data for 172 cycles spread over 250 000 cycles of the 569-s modulation).

4. Discussion

To determine the period evolution of short period AM CVn stars a dedicated long-term campaign is needed in order to eliminate aliases and cycle count uncertainties. After two years of observations of ES Cet, the O–C diagram is starting to show a slight upwards curvature (indicating a increase in period) and we derive a (tentative) value for $\dot{P} \sim 1.6 \times 10^{-11}$, or the equivalent $\dot{\nu} \sim -4 \times 10^{-17}$ Hz s^{-1}. Insecure though it is, it may be the first detection of a $\dot{P}$ in an AM CVn system.

Acknowledgments

PAW's research if funded by a strategic grant from the University of Cape Town and by funds from the National Research Foundation. BW's research is funded by the University of Cape Town. We kindly thank the HET Board for granting observing time.

References

Abbott, T.M.C., Robinson, E.L., Hill, G.J., et al. 1992, *ApJ*, **399**, 680.

Burwitz, V., Reinsch, K. 2001, In X-ray astronomy: stellar endpoints, AGN, and the diffuse X-ray background, eds. N.E. White, G. Malaguti, G.G.C. Palumbo. AIP Conf. Proc., **599**, 522.

Chornock, R., Filippenko, A.V. 2003, *IAUC*, **8084**, 3.

Cropper, M., Harrop-Allin, M.K., Mason, K.O., et al. 1998, *MNRAS*, **293**, 57L.

Hakala, P., Ramsay, G., Wu, K., et al. 2003, *MNRAS*, **343**, 10L.

Israel, G.L., Hummel, W., Covino, S., et al. 2002, *A&A*, **368**, 13L.

Marsh, T.R., Horne, K., Rosen, S. 1991, *ApJ*, **366**, 535.

Nather, R.E., Robinson, E.L., Stover, R.J. 1981, *ApJ*, **244**, 269.

O'Donoghue, D. 1995, *Baltic Astr.*, **4**, 519.

O'Donoghue, D., Kilkenny, D., Chen, A., et al. 1994, *MNRAS*, **271**, 910.

Patterson, J., Fried, R.E., Rea, R., et al. 2002, *PASP*, **114**, 65.

Patterson, J., Kemp, J., Chambrook, A., et al. 1997, *PASP*, **109**, 1100.

Patterson, J., Wlaker, S., Kemp, J., et al. 2000, *PASP*, **112**, 625.

Ramsay, G., Hakala, P., Cropper, M. 2002, *MNRAS*, **332**, 7L.

Ruiz, M.T., Rojo, P.M., Garay, G., Maza, J. 2001, *ApJ*, **552**, 679.

Schwarz, H.E., van Amerongen, S., Heemskerk, M.H.M., et al. 1988, *A&A*, **202**, 16L.

Skillman, D.R., Patterson, J., Kemp, J., et al. 1999, *PASP*, **111**, 1281.

Solheim, J.-E., Provencal, J.L., Bradley, P.A., et al. 1998, *A&A*, **332**, 939.

Steeghs, D. 2003, Workshop on Ultracompact Binaries, Santa Barbara. See `http://online.kitp.ucsb.edu/online/ultra_c03/steeghs/`

Strohmayer, T.E. 2002, *ApJ*, **581**, 577.

Strohmayer, T.E. 2003, *ApJ*, **593**, 39L.

van Amerongen, S., van Paradijs, J. 1989, *A&A*, **219**, 195.

Warner, B. 1995, Ap&SS, **225**, 249.

Warner, B. 2005, these proceedings.

Warner, B., Woudt, P.A. 2002, *PASP*, **114**, 129.

Wood, M.A., Casey, M.J., Garnavich, P.M., et al. 2002, *MNRAS*, **334**, 87.

Wood, M.A., Winget, D.E., Nather, R.E., et al. 1987, *ApJ*, **313**, 757.

Wood-Vasey, W.M., Aldering, G., Nugent, P., Li, K. 2003, *IAUC*, **8077**, 1.

Woudt, P.A., Warner, B. 2002, *MNRAS*, **328**, 159.

Woudt, P.A., Warner, B. 2003a, *MNRAS*, **345**, 1266.

Woudt, P.A., Warner, B. 2003b, *IAUC*, **8085**, 3.

PHOTOMETRIC OBSERVATIONS OF CATACLYSMIC VARIABLES WITH THE 1.5-M TELESCOPE AT THE TÜBITAK NATIONAL OBSERVATORY (TUG): V2275 CYG, RW UMI, PX AND AND FO PER

Ş. Balman,[1] A. Yılmaz,[1] T. Ak,[2] T. Saygaç,[2] H. Esenoğlu,[2] A. Retter,[3] Y. Lipkin,[4] U. Kızıloğlu,[1] A. Bianchini,[5] S. Aliş[2,6]

[1]*Department of Physics, Middle East Technical University, Ankara, 06531, Turkey*
solen@astroa.physics.metu.edu.tr

[2]*Istanbul University, Dept. of Astronomy and Space Sciences, Istanbul, 34452, Turkey*

[3]*School of Physics, University of Sydney, 2006, Australia*

[4]*School of Physics and Astronomy, and the Wise Observatory, Tel-Aviv University, Israel*

[5]*Departmet of Physics, University of Padova, Vicolo dell'Osservatorio 5, 35122, Padova, Italy*

[6]*Eyüboğlu Educational Institutions, Eyüboğlu Twin Observatories, 34762, Ümraniye-Istanbul, Turkey*

Abstract We present the light curve and time series analysis of classical novae and dwarf novae systems monitored/observed with the 1.5 m Russian-Turkish Joint telescope (RTT150) at the TÜBITAK (The Scientific and Technical Research Council of Turkey) National Observatory in Antalya, Turkey. As part of a large program on CCD photometry of Cataclysmic Variables, V2275 Cyg (N 2001 No.2), RW UMi, FO Per and PX And were observed for a total of about 25 nights. The results on V2275 Cyg show that the system has a period of 0.463±0.014 or the 1-d alias 0.316±0.007 with a wide eclipse-like pattern (IAUC 8074). The power spectral analysis of the data on RW UMi reveal possible periodicities at around several frequencies (eg., 14.4, 16.7, 19, 30, 39, 46, 68, 108 in cycles per day) that could be interpreted as the binary period, spin period of the white dwarf and/or orbital sidebands of the system. We find that the radial velocity profiles of H-alpha lines and the photometry of the Dwarf Nova FO Per indicates the possible presence of a period around 0.1828 d.

Keywords: accretion disks, cataclysmic variables, close binaries, novae, white dwarfs

E.M. Sion, S. Vennes and H.L. Shipman (eds.), White Dwarfs: Cosmological and Galactic Probes, 233-242.
© 2005 *Springer.*

1. V2275 Cyg (N 2001 No.2): Discovery of Variations

V2275 Cyg (Nova Cyg 2001 No.2) was discovered at magnitude 8.8 on 2001 August (Nakamura et al. 2001, Nakano et al. 2001). At later stages, high energy coronal lines were found to dominate the spectrum with [Si X], [Si IX] and [Al IX]. The nova was found to belong to the "He/N" subclass of novae defined by Williams (1992), because of the broad lines of H, He and Ne in its spectrum (Kiss et al. 2002). The early photometry reveals t_2=2.9±0.5 d and t_3=7±1 d. The distance is 3-8 kpc (Kiss et al. 2002).

Our data were obtained with the Ap47p CCD (1024×1024 pixels with 13 microns/pix) at TÜBITAK National Observatory (TUG) for 8 nights : 2002 June 10 and 12, October 1 to 5, and December 1 and 22; each 90 sec exposure and standard R filters were used. A total of 552 frames were accumulated and reduced within standard procedures using dark current frames and dome flat-field frames. After the raw data were cleaned and calibrated, the instrumental magnitudes of the nova and four nearby comparison stars were derived by PSF fitting algorithm of DAOPHOT (Stetson 1987) and ALLSTAR within MIDAS software package. The calibrated magnitude of the nova varied between 15.1 and 16.2 mags. Figure 1a shows the image of the field of the nova and Figure 1b shows the light curve of the classical nova. The relative magnitudes were calculated using four comparison stars and averaged to reduce scintillation effects. Time series analysis has been performed using Scargle algorithm, Multi Harmonic Analysis of variance and Discrete Fourier Transform algorithms (see Figure 2). We detect large variations in the light curve of this classical nova similar to eclipsing binary systems with a period of either 0.463±0.014 day or 1-day alias at 0.316±0.007 day, reported also by our group in IAUC 8074 (Balman et al. 2003). Ephemeris for the period detected at 0.463 day and 0.316 day using a gaussian function to fit the data set are :

$$\text{MAX (HJD)} = 2452436.5087(\pm 0.0147) + 0.31620(\pm 0.007)E$$
$$\text{MAX (HJD)} = 2452436.5173(\pm 0.0112) + 0.46267(\pm 0.014)E$$

The semi amplitude of variations are 0.487±0.012 (0.463 d) and 0.447±0.013 (0.316 d) respectively obtained from the fits with a sine function. The follow-up observations of this nova is being carried out by our group at TUG in order to predict the correct period, derive color variations and other possible periodicities (see also Discussion and Conclusions).

2. RW UMi (1956): An Intermediate Polar ??

RW UMi is a Galactic classical nova system that had an outburst in 1956. Since then the system has been observed to have a bright disk and the optical light curve shows several indefinite periodicities : 117±5 min (Szkody et al. 1989), 113±10 min (Howell et al. 1991), 0.05912±0.00015 d (1.4 hrs- Retter & Lipkin 2001). The system indicates a 2 hrs-1.7 hrs main period and several

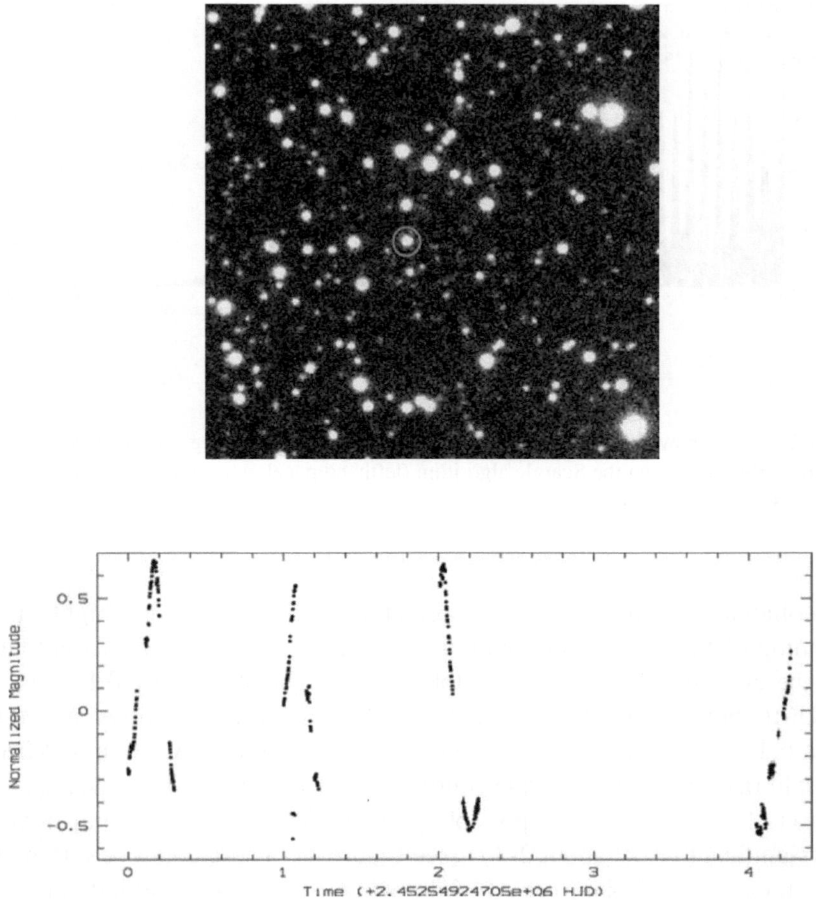

Figure 1. (a) The field of the classical nova (V2275 Cyg). North is up and West is to the right. The nova is circled. The image was obtained on 2002 October 02 with a standard R filter at TUG. (b) The light curve of V2275 Cyg (2001) obtained using the Johnson R Band Filter with the 1.5m telescope at TUG during the nights of 01-05 October 2002. Errors are indicated on the data points.

quasiperiodic oscillations at higher frequencies $\sim$ five different periods that *could* be interpreted as spin and sideband frequencies from an Intermediate Polar system (Bianchini et al. 2003).

We have been observing RW UMi with the 1.5 m Telescope at TUG using the imaging CCD : 2048$\times$2048 with 15microns/pix (Liquid Nitrogen cooling system). We have compiled seven nights in 2001 : 21,23 June (R Filter); 23-25 July (R Filter); 15 August (R Filter) and 16 September (R Filter). We have

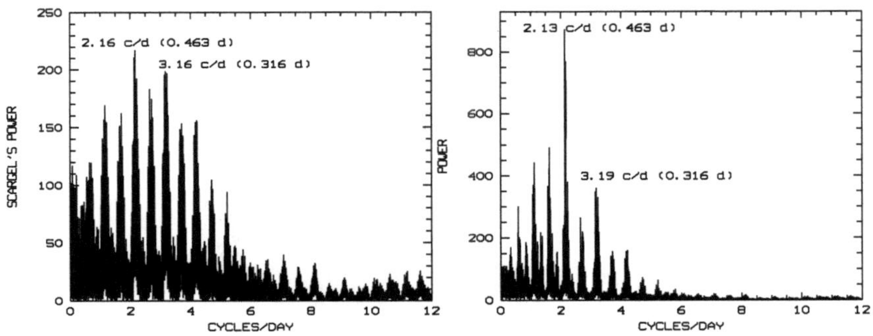

Figure 2. The power spectrum of V2275 Cyg obtained performing a transform of the entire time series data using the Scargle algorithm (lefthand panel) and Multiharmonic analysis-2 harmonics (righthand panel).

also collaborated with the WISE observatory in Israel and got 5 nights at the beginning of September 2001 with the 1 m telescope. WISE data were taken with the clear filter. We present a total of 12 nights of data (440 frames) for RW UMi obtained in 2001 (see Figure 3). In general, observations ranged between 6-1.5 hrs each night and exposures were 2 min (TUG)-4 min (WISE). The calibrated magnitude of the source was 18.7-19.1 . The power spectral analysis of RW UMi reveals possible periodicities at around several frequencies (significant peaks about 2-3σ) (see Figure 4). These could be interpreted as the binary period, spin period of the white dwarf and/or orbital sidebands of the system. The errors on the frequencies are about $\pm$ 0.5 c/d. In order to remove the aliasing problem, more data is necessary. We have observed the source in 2003 with TUG and have 5 nights of data (2 of which is from WISE) in hope to recover the intriguing puzzels of this system.

3. A Puzelling SW Sex Star: PX And (PG0027+260)

PX And is a complicated SW Sex-type object that shows shallow eclipses with highly variable depth (Thorstensen et al., 1991), negative superhumps with a period of about 0d.142 and a disk precession period of about 4d.8 (Stanishev et al., 2002). Detailed investigations of PX And have also been made by Hellier and Robinson (1994) and Still et al. (1995).

We began a project including photometric observations of several SW Sex stars for a better understanding of this phenomenon. Photometric CCD obser-

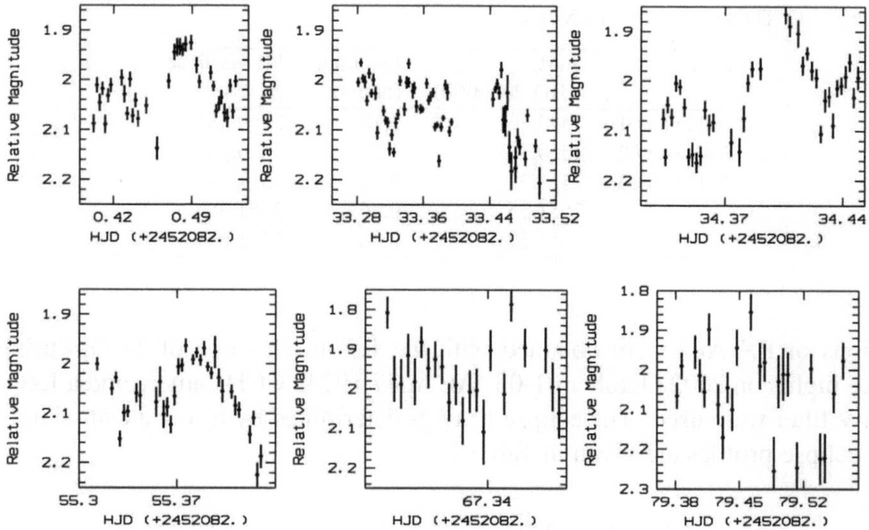

Figure 3. The light curves of RW UMi. First four panels are obtained at TUG with the 1.5 m Telescope using Johnson-R filter during 2001 July 23-25 and 2001 August 15. The last two panels are part of the observations obtained from the WISE observatory 1.0 m telescope in 2001 September (clear filter).

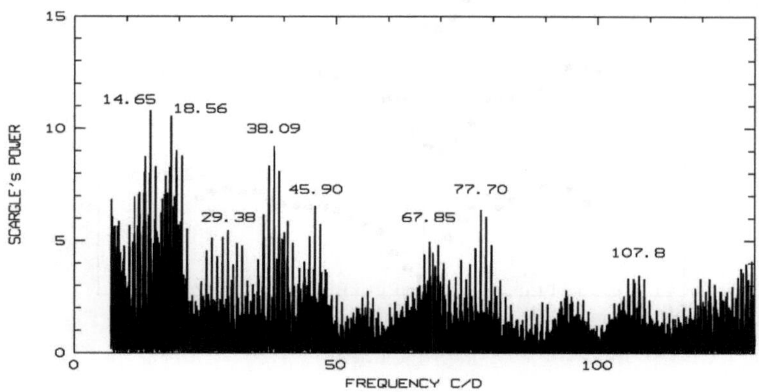

Figure 4. The power spectrum of RW UMi. Once the data is prewhitened from 8 definite periods that exists in the window transform, the leftover prominents peaks are 14.4, 19.3 and 39.1 c/d . A discrete fourier transform is used for the time series analysis.

Table 1. CCD Observations of PX And

Date	Start (HJD-2450000)	Duration (hours)	Mid-Eclipse (HJD)
2002 Oct 01	2549.46	1.18	2549.481
2002 Oct 02	2550.34	1.20	2550.360
2002 Oct 03	2551.22	0.72	2551.238
	2551.36	0.96	2551.384
	2551.52	0.96	2551.533

vations of PX And were obtained with the 1.5 m telescope of TUG, during three nights on 2002 October 01-03. An Ap47 10242 CCD camera and a Johnson V filter were used. The eclipse timings determined by fitting a Gaussian to the eclipse profiles are given in Table 1.

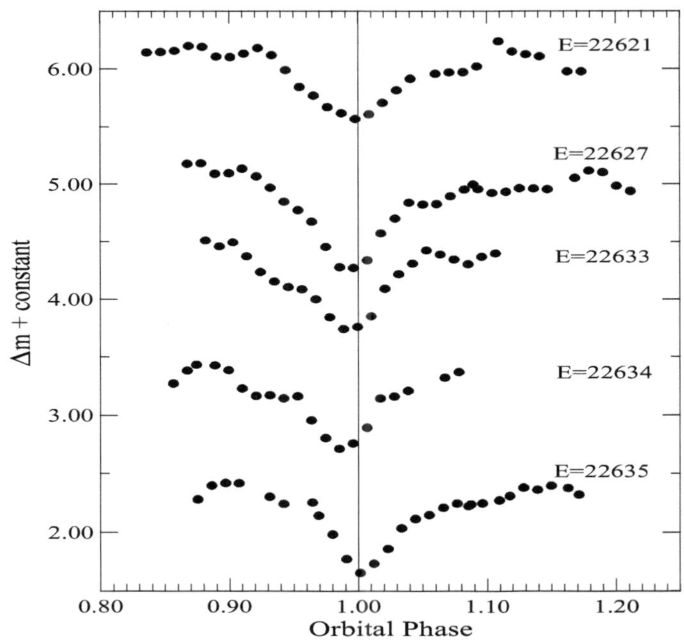

Figure 5. The V band eclipse profiles of PX And, relative to five comparison stars in the field, obtained with TUG 1.5 m Telescope. Mean errors are 0.003 mag.

Combining the above values with the eclipse timings published by Yong et al. (1990), Hellier and Robinson (1994), Shakhvskoy et al. (1995) and Stanishev et al. (2002), we determined the following orbital ephemeris:

$$\text{MIN (HJD)} = 2449238.8374(9) + 0.146352724(6)\ E$$

Figure 5 shows the eclipse profiles. All the folded profiles were normalized using the out of eclipse magnitudes (average of 5 exposures at about maxima). We used mean differential magnitudes obtained from 5 comparison stars in the field (Henden and Honeycutt, 1995). The eclipse profiles reveal variable eclipse depth as was predicted : cycle number 22627 is about 0.87 mag though others are measured in a range 0.61-0.75 mag. Profiles show post-eclipse humps, also. The further observations and detailed study on the physical model of this object is in progress.

4. An Attempt to Recover the Period of FO Per

FO Per was discovered and classified by Hoffmeister (1967) as a dwarf nova. It was referenced as U Gem type star in Morris et al. (1987). We found that the previous spectral and photometric studies of this source were inadequate and started a project at the Asiago Observatory (Italy) and follow-up observations at TUG to determine the spectral characteristics and the binary period of this source.

We attempted to detect radial velocity variations of the H-alpha emission of FO Per in the intermediate resolution spectra taken during four observation runs, in 1990, 1994, 1995 and 1998, carried out with the 1.82-m telescope of the Asiago Observatory (Italy) equipt with a B&C grating spectrograph (TH 7882 CCD, grating-1200 gr/mm, dispertion 42 Å/mm or 1.0 Å/pixel, and resolution 1.9 Å). We have measured from the first results of the spectroscopic data (70 frames) a rather irregular radial velocity curve suggesting a rough orbital period of 0.18248 d (see Figure 6a). In order to derive the period of the binary system, we also carried out photometric observations. Six observation runs are done in 1996 (1.82 m telescope Asiago Observatory, Italy), the fourier transform is displayed in Figure 6b. In addition, we have obtained observations at TUG 1.5 m Telescope (see Figure 7a,b): Three nights in December 2002, three nights in March of 2003. The B, V, R filters were used with the (AP47p and imaging CCDs). Our analysis and the monitoring of the dwarf novae is in progress. The photometric light curve indicates short term variability. The possible period of FO Per falls in the expected range of dwarf novae periods among the U Gem stars. The single peak shaped H-alpha profile, large absorption on the base of H-alpha and the changing intensity of large absorption along the phase of the system is interesting. This absorption component was not recovered in the previous spectral analysis (Bruch 1989, Schimpke & Bruch 1992), however our data cover several different phases of the star. The system will be understood better after the reduction of the photometric data and the rest of the spectra.

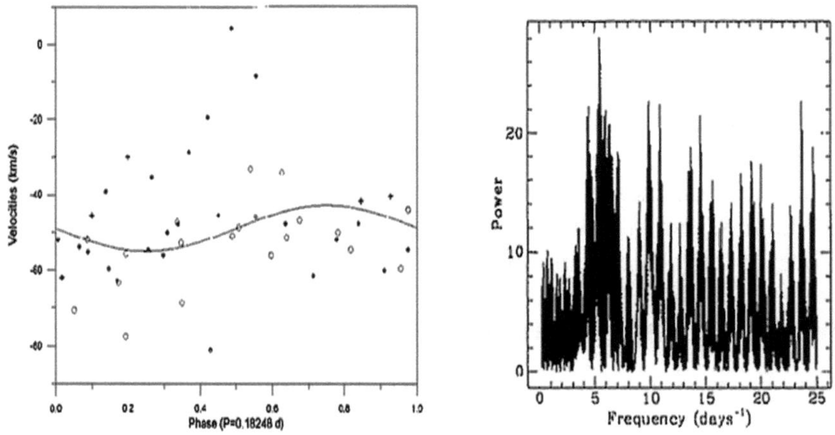

Figure 6. (a) The left-hand panel shows the folded H-alpha radial velocities on the 0.18248 d period. The data is obtained from the Asiago Observatory and contains quiescence and outburst observations. (b) The right-hand panel shows a reduced spectrum for only wavelength calibration in order to calculate radial velocities of H-alpha profiles together with identification of continuum level under IRAF.

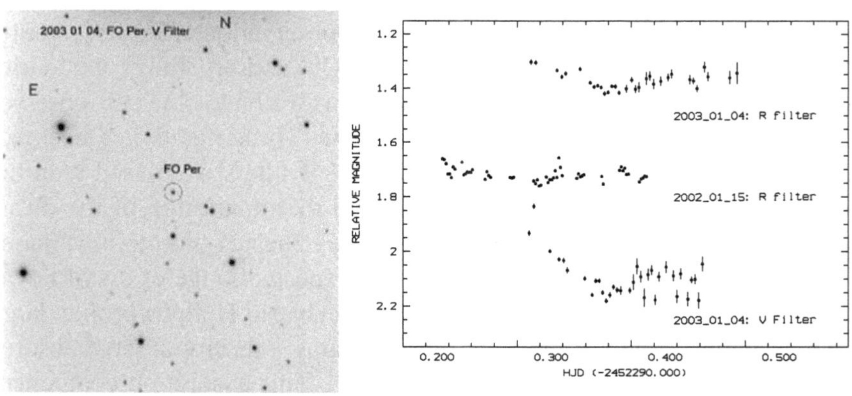

Figure 7. (a) The left-hand panel is the field of FO Per. The source is indicated with a circle. The observation was obtained with a standard V filter using the imaging CCD. (b) The right-hand panel shows the light curve obtained with TUG 1.5 m telescope on 2003 January 04 and 2002 January 15.

5. Discussion and Conclusions

We have discovered deep modulations in the light curve of the classical nova 2001 No.2, V2275 Cyg. Since it is the strongest period in the power spectrum of the light curve this could be the binary period of the system. The cause of modulations could be an eclipse and/or the aspect variations of the secondary due to heating from the hot WD. At the moment (Jan 2004) we have acculumated enough data and by appropriate removal of the rednoise from the power spectrum, we can distingish the two aliases and the system show different variations as well. These will be presented in Balman et al. (2004) in preperation.

RW UMi shows several interesting peaks in its power spectrum that are at 2-3 σ confidence level. The total data we have on the object is not long enough (can not remove aliasing) to conclude on the real periods since the modulation depths are by far smaller compared with V2275 Cyg. The best candidates for the periodicities from the system are 14.4, 19.3 and 39.1 c/d with an erorr of about 0.5 c/d. It is likely that this system is an intermediate polar candidate.

The modulation in the radial velocities of FO Per expected to be due to the variations of the hot spot on the disk are small and scattered in a large interval derived from our medium dispersion spectra. The small modulation and large scattering indicates low inclination for FO Per. The narrow H-alpha profiles in quiescence are, also, an indicator of the low inclination of the system. The photometry of FO Per with TUG 1.5 m telescope suggest existence of variations, the analysis is in progress.

We presently have three programs at the 1.5-m telescope of TUG observatory on the monitoring of the light curves and performing time series analysis for old classical novae, classical novae in outburst, dwarf novae and SW Sex stars. A TFOSC spectrometer will be avaliable for GO use starting by the end of this year where simultaneous spectra and photometry will be obtained with high sensitivity and resolution. TUG observatory has another 40 cm telescope at the cite together with a robotic telescope that will be installed at the mountain this summer for the ROTSE IIId project where serendipitous discovery of novae and prompt follow-ups after outburst can be conveniently performed. The details could be find at "http://www.tug.tubitak.gov.tr".

Acknowledgments

We thank the TÜBITAK National Observatory Institute for providing travel support to the observatory site and to Z. Aslan, M. Parmaksızoğlu, I. Hamitov and K. Uluç for various help provided during observations.

References

Balman S., Yılmaz A., Retter, A., Ak, T., Saygaç, T., Esenoğlu, H. and Aslan, Z., *IAUC*, **8074**.

Bianchini, A., Tappert, C., Cantera, R., Tamburini, F., Osborne, H., and Cantrell, K. 2003, *PASP*, **115**, 811.

Bruch, A. 1989, *A&A Suppl. Ser.*, **78**, 145.

Hellier, C., Robinson, E.L. 1994, *ApJ.*, **431**, L107.

Henden, A.A., Honeycutt, R.K. 1995, *PASP*, **107**, 324.

Hoffmeister, C. 1967, *Astron. Nachr.*, **289**, 205.

Howell, S., Dobrzycka, D., Szkody, P., and Kreidl, T. 1991, *PASP*, **103**, 300.

Kiss, L. L., Gogh, N., Vinko, J., Furesz, G., Csak, B., DeBond, H., Thomson, J. R. and Derekas, A. 2002, *A&A*, **384**, 982.

Morris, S.L., Schmidt, G.D., Liebert, J., et al. 1987, *ApJ*, **314**, 641.

Nakamura, A., Tago, A., and Abe, H. 2001, *IAUC*, **7686**, 2.

Nakano, S., Hatayama, K., Kato, T., et al. 2001, *IAUC*, **7687**.

Retter, A. and Lipkin, Y. 2001, *A&A*, **365**, 508.

Schimpke, T. and Bruch, A. 1992, *A&A*, **266**, 225.

Shakhovskoy, N.M., Kolesnikov, S.V., Andronov, I.L. 1995, in Cataclysmic Variables, Proc. of the cenference held in Abano Terme, Italy, 20-24 June 1994 (Dordrecht Kluwer Academic Publishers), ed. A. Bianchini, M. Della Valle, & M. Orio, *Astrophys. Space Sci. Libr.*, **205**, 187.

Stanishev, V., Kraicheva, Z., Boffin, H.M.J, and Genkov, V. 2002, *A&A*, **394**, 625.

Still, M.D., Dhillon, V.S., and Jones, D.H.P. 1995, *MNRAS*, **273**, 863.

Szkody, P., Howell, S., Mateo, M., and Kreidl, T. J. 1989, *PASP*, **101**, 899.

Thorstensen, J.R., Ringwald, F.A., Wade, R.A., Schmidt, G.D., and Norsworthy, J.E. 1991, *AJ*, **102**, 272.

Williams, R. E. 1992, *AJ*, **104**, 725.

Yong, L., Zhaoji, J., Jiangsheng, C., and Mingzhi, W., 1990, *IBVS*, **3434**.

THE POSSIBLE IDENTIFICATION OF TWO HIBERNATING NOVAE

Adéla Kawka
Division of Science and Engineering
Murdoch University, WA, 6150, Australia
kawka@maths.anu.edu.au

Stéphane Vennes
Department of Mathematics
Australian National University, ACT, 0200, Australia
vennes@maths.anu.edu.au

Abstract We have obtained spectroscopic and photometric measurements of the close DA plus dMe binaries BPM 71214 and EC 13471−1258, and established their orbital parameters. We measured an orbital period of 0.201626 day for BPM 71214 and 0.15074 day for EC 13471−1258. The light curves show the presence of ellipsoidal variations in both systems with EC 13471−1258 also being an eclipsing binary. Detailed modeling of the spectroscopic and photometric data (R and I for BPM 71214, and B and R for EC 13471−1258) results in mass ratios in the range 0.5 to 0.7 and shows that the red dwarfs in both systems are nearly filling their Roche lobes. Because there is no evidence of mass transfer in the systems, we conclude that BPM 71214 and EC 13471−1258 are possibly hibernating novae.

Keywords: close binaries, novae

1. Introduction

The hibernating nova scenario was introduced to explain the low rate of observed novae in our Galaxy compared to the predicted rate of nova outbursts from nova theory (Patterson 1984, Shara et al. 1986). The hibernating nova theory would allow novae to hide in the local population of white dwarfs.

Close binary systems can evolve into novae via the accumulation of mass on the white dwarf surface through non-conservative mass transfer which will lead to a thermonuclear runaway (TNR). Following the TNR, the secondary continues to transfer mass onto the white dwarf, which is induced by the irradiation

E.M. Sion, S. Vennes and H.L. Shipman (eds.), White Dwarfs: Cosmological and Galactic Probes, 243-249.
© 2005 *Springer*.

of the red dwarf surface by the hot white dwarf. This mass transfer following a nova outburst has been observed in a number of nova systems (Warner 1995). As the white dwarf cools, the secondary contracts until it underfills its Roche lobe allowing the mass transfer to become very low or cease completely. The system is now in hibernation and after 10^3 to 10^6 years the system will again initiate mass transfer following orbital momentum loss through gravitational radiation or magnetic breaking. The general scenario is described in more detail in Shara (1989).

In general, cataclysmic variables appear to evolve through cycles. King et al. (1995) proposed a mass-transfer cycle which is driven by the irradiation of the secondary, and which determines the accretion rate. This allows high-states (where irradiation expands the secondary) and low-states (where the secondary contracts and mass transfer rate decreases). In this scenario, the cataclysmic variable (CV) spends similar amounts of time in the high- and low-accretion states, and the probability that a CV will become a nova is highest during the high mass transfer rate. Similarly, Prialnik & Kovetz (1995) calculated multi-cycle nova evolutionary models for systems with white dwarf masses ranging from 0.65 to $1.4M_{\odot}$. They show that these systems go through nova outbursts every 100 to 10^6 years. They also found that the accretion rates vary considerably from system to system implying that nova systems can go through periods of hibernation. In addition, all CVs must undergo nova eruptions, which implies that all known CVs have undergone a nova eruption in the past or will undergo a nova eruption in the future (Warner 1995).

Many of the aspects related to the evolution of CVs are still being debated, such as the role played by irradiation in the mass transfer rate (Schreiber et al. 2000), and the effect of nova outbursts on the mass transfer rate which may explain the wide spread of mass transfer rates for systems above the period gap (Kolb et al. 2001). We present a study of two systems which are most likely in the hibernation phase of their mass transfer cycle.

EC 13471−1258 was observed as part of the Edinburgh-Cape Blue Object Survey. Kilkenny et al. (1997) reported the object as variable and classified it as a DA plus dMe eclipsing binary with an orbital period of 3 hr 37 minutes. Kawka et al. (2002) observed this object as part of a study of 4 close binaries. More recently, O'Donoghue et al. (2003) reported their observations of this system. BPM 71214 has been classified as a post-common-envelope binary by Hillwig, Honeycutt & Robertson (2000) and as a pre-CV by Marsh (2000), or a possible hibernating nova by Livio & Shara (1987) and Sarna, Marks & Smith (1995). BPM 71214 was observed spectroscopically by Kawka et al. (2002) who suggested that this system, like EC 13471−1258, is a hibernating nova rather than a pre-CV. The spectroscopic observations were followed up with photometry, where ellipsoidal variations were observed confirming the nature of BPM 71214 (Kawka & Vennes 2003). We summarize the work done in

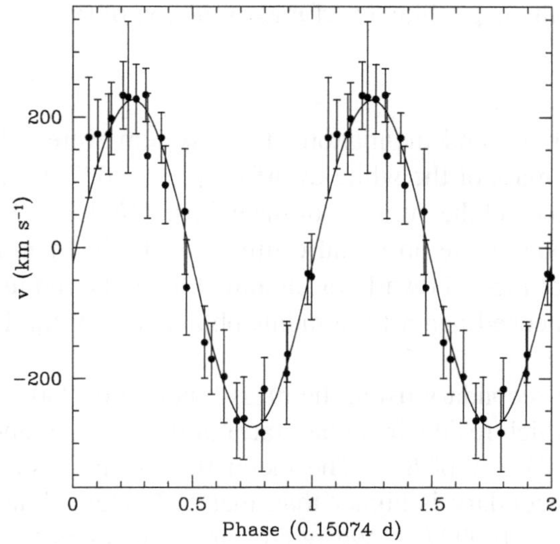

Figure 1. Radial velocities traced by absorption features in the wavelength region 6240 to 6540 Å for EC 13471−1258 and folded on the orbital period, compared to the best sinusoidal fit, which has a semi-amplitude of 251 ± 12 km s^{-1} and a systemic offset of -24 ± 8 km s^{-1}.

Kawka et al. (2002) and Kawka & Vennes (2003), and we report on additional work done on the close binary EC 13471−1258.

2. EC 13471−1258

Orbital parameters

The orbital period of EC 13471−1258 is 0.15074 ± 0.00004 days (Kawka et al. 2002). Using Hα emission Kawka et al. (2002) measured a red dwarf velocity semi-amplitude of 241.0 ± 8.1 km s^{-1}, however using absorption features of the red dwarf O'Donoghue et al. (2003) obtained a semi-amplitude of 266 ± 6 km s^{-1}. We remeasured our radial velocities of the red dwarf by cross-correlating the absorption features of EC 13471−1258 in the range of 6240 to 6540 Å against two M-type spectra (GL 190: M3.5 and GL 250B: M2) using FXCOR in IRAF. Prior to cross-correlation, both template spectra were smoothed to the rotational velocity (140 km s^{-1}) of the secondary. The velocities were phased with the orbital period, and fitted with a sine curve using a least square fit to obtain $K_{sec} = 246 \pm 14$ km s^{-1} and a systemic velocity of $\gamma = -19 \pm 10$ km s^{-1} when the template star GL 190 was used, and $K_{sec} = 251 \pm 12$ km s^{-1} and $\gamma = -24 \pm 8$ km s^{-1} when the template star GL 250B was used. Figure 1 shows the radial velocity measurements that were determined using the template star GL 250B. Combining our velocity semi-amplitude determined from absorption profiles with the one obtained from Hα emission lines, the weighted average of the velocity semi-amplitude is 245 ± 6 km s^{-1}, which results in a white dwarf mass function of $0.230 \pm 0.017 M_{\odot}$.

Light curves

The red dwarf mass and inclination of the system were calculated using the mass function, mass of the white dwarf ($M_{WD} = 0.77 M_{\odot}$: Kawka et al. 2002), orbital period and the duration of the eclipse (15.4 ± 1.0 min) as inputs into an iterative code where an initial estimate of the inclination of 80° was made. The radius of the red dwarf was assumed to be the radius of the Roche lobe (which was checked against the radius obtained from the FWHM of Hα emission).

The mass of the secondary using the mass function ($0.230 \pm 0.017 M_{\odot}$) is $0.55 \pm 0.11 M_{\odot}$, which results in a mass ratio of 0.72 ± 0.10 and a separation of the binary of $1.31 \pm 0.05 R_{\odot}$. The inclination of the system is $74 \pm 2°$. The mass of the secondary is higher than that of O'Donoghue et al. (2003) who obtained $0.41 \pm 0.09 M_{\odot}$. We redid our calculations by replacing our $K_{RD} = 245$ km s^{-1} with $K_{RD} = 266$ km s^{-1} from O'Donoghue et al. (2003) to obtain a mass of the secondary of $0.41 \pm 0.09 M_{\odot}$ with a binary separation of $1.26 \pm 0.04 R_{\odot}$. The inclination would be $76 \pm 2°$, which is consistent with the inclination determined with the lower mass function.

The radius of the Roche lobe was calculated to be $0.46 \pm 0.02 R_{\odot}$ which is in agreement with the secondary radius measured from the FWHM of the $H\alpha$ emission. This implies that the secondary is just underfilling its Roche lobe, since there does not appear to be any evidence of mass transfer. Using the mass-radius relation for M-dwarfs, the expected radius for an M-dwarf with a mass of $0.55 M_{\odot}$ is $0.57 R_{\odot}$, which contradicts the measured value, however a comparison to other M-dwarfs of similar mass shows that the radius is not unreasonably too small, and that it is within the scatter of values (Caillault & Patterson 1990, Clemens et al. 1998)

We have compared the light curve of EC 13471−1258 to model light curves which were calculated using the Wilson-Devinney code (WD: Wilson 1979, 1990). We have assumed a detached system (MODE = 2), blackbody source spectra with no spots present on the surface of the primary or the secondary, and reflection was allowed with $albedos = 1$. The white dwarf mass was fixed at $0.77 M_{\odot}$, the inclination at 73.5° and the white dwarf temperature at 14085 K. The radius of the white dwarf was constrained by the spectroscopic surface gravity and the mass-radius relations of Wood (1995). The temperature of the secondary was varied until the depth of the eclipse was satisfied. The radius of the secondary was varied until the amplitude of the ellipsoidal variations was satisfied.

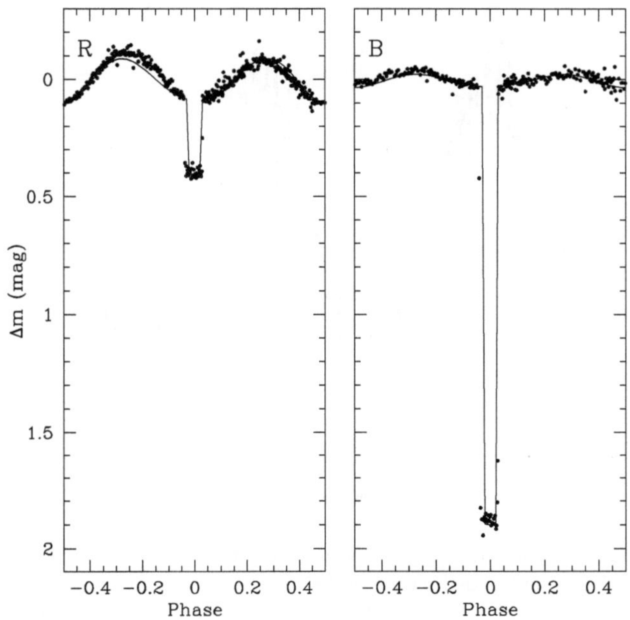

Figure 2. Photometry of EC 13471−1258 in R and B showing ellipsoidal variations and the eclipse. The depth of the eclipse is 0.3 magnitudes in R and 1.85 magnitudes in B with a duration of 15.4 minutes. The semi-amplitudes of the ellipsoidal variations are 0.097 and 0.025 magnitudes in R and B, respectively. The light curves are compared to model light curves (solid black line). There is no evidence of illumination of the red dwarf by the white dwarf.

The mass of the secondary was fixed at $0.55M_\odot$. However, model light curves were also computed using a secondary mass of $0.42M_\odot$. Note, that when the mass of the secondary is decreased the separation of the binary must also decrease, and, therefore, the mass of the secondary has little effect on the resulting light curve unless strict mass-radius relations are kept.

First, the R light curve was fitted, and $T_{RD} = 2790$ K and $R_{RD} = 0.44R_\odot$ provided the best fit. These parameters were then adopted for the B light curve, but the secondary temperature had to be increased to 2980 K. Since the temperature of the M-dwarf cannot be well represented by a blackbody, the secondary temperature of 2900 ± 150 K is only an estimate. Figure 2 shows the R and B light curves compared to best fit model light curves.

3. BPM 71214

Orbital parameters

The orbital period of BPM 71214 is 0.201626 ± 0.000004 days (Kawka & Vennes 2003). Using Hα emission Kawka & Vennes (2003) measured a red

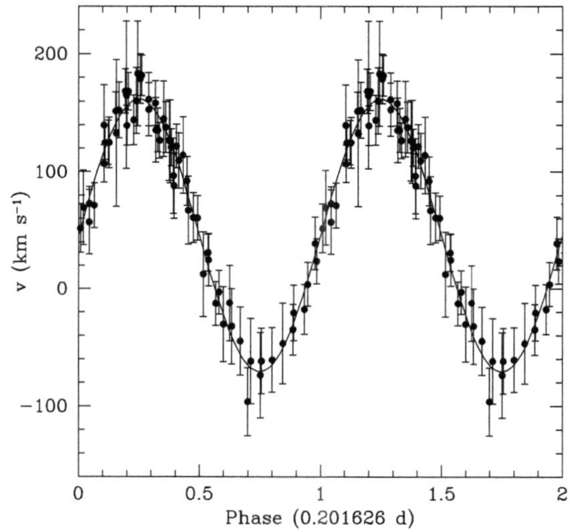

Figure 3. Radial velocities traced by absorption features in the wavelength region 6240 to 6540 Å for BPM 71214 and folded on the orbital period, compared to the best sinusoidal fit, which has a semi-amplitude of 116.2 ± 2.5 km s^{-1} and a systemic offset of 45.8 ± 1.8 km s^{-1}.

dwarf velocity semi-amplitude of 121.4 ± 1.6 km s^{-1} and a systemic velocity of 48.7 ± 1.1 km s^{-1}. We have remeasured the radial velocities of the red dwarf by cross-correlating the absorption feature of BPM 71214 in the range 6240 to 6540 Å against the M-dwarf, GL 250B using FXCOR in IRAF. The velocities were phased with the orbital period, and fitted with a sine curve using a least square fit to obtain $K_{sec} = 116.2 \pm 2.5$ km s^{-1} and $\gamma = 45.8 \pm 1.8$ km s^{-1}. The radial velocity measurements are shown in Figure 3.

Light curves

Photometric observations in R and I of BPM 71214 were obtained by Kawka & Vennes (2003). These observations were phased with the orbital period and the resulting light curves were compared to model light curves, which were calculated using the WD code. Kawka & Vennes (2003) found that the best fit to their light curves occurred when the Roche lobe of the secondary was filled (at $q \sim 0.7$), suggesting that the secondary should be transferring mass onto the white dwarf. However, no evidence of any mass transfer was found, implying that the secondary in BPM 71214, like in EC 13471−1258, must be just underfilling its Roche lobe.

4. Discussion

The orbital parameters and photometry for EC 13471−1258 and BPM 71214 suggest that both systems are close to filling their Roche lobe. Therefore, these systems may either be post-common envelope binaries just prior to mass transfer, or they may be hibernating CVs. Schreiber & Gänsicke (2003) calculated the fractional PCEB life-time that a selected number of systems have passed through. For EC 13471−1258 and BPM 71214 they found that both systems, assuming that they are PCEB, have passed through more than 95% of their life-time. It is more likely that we are seeing two systems that are in hibernation rather than systems which are just about to begin mass transfer. The theory of novae predicts that once a system begins mass transfer, the binary will begin a cycle of mass transfer, which will lead to a nova outburst every 100 to 10^6 years. Between outbursts the system can enter hibernation for extended periods of time. These systems will appear as detached white dwarf plus M-dwarf binaries that are just underfilling their Roche lobe. BPM 71214 and EC 13471−1258 are two such binaries, and therefore should be considered as hibernating novae.

References

Caillault, J.-P., & Patterson, J. 1990, *AJ*, **100**, 825.

Clemens, J. C., Reid, I. N., Gizic, J. E., & O'Brien, M. S. 1998, *ApJ*, **496**, 352.

Hillwig, T. C., Honeycutt, R. K., & Robertson, J. W. 2000, *AJ*, **120**, 1113.

Kawka, A., Vennes, S., Koch, R., Williams, A. 2002, *AJ*, **124**, 2853.

Kawka, A., & Vennes, S. 2003, *AJ*, **125**, 1444.

Kilkenny, D., O'Donoghue, D., Koen, C., Stobie, R.S., & Chen, A. 1997, *MNRAS*, **287**, 867.

King, A.R., Frank, J., Kolb, U., & Ritter, H. 1995, *ApJ*, **444**, L37.

Kolb, U., Rappaport, S., Schenker, K., & Howell, S. 2001, *ApJ*, **563**, 958.

Marsh, T.R. 2000, *NewA Rev*, **44**, 119.

O'Donoghue, D., Koen, C., Kilkenny, D., Stobie, R.S., Koester, D., Bessell, M.S., Hambly, N., & MacGillivray, H. 2003, *MNRAS*, **345**, 506.

Patterson, J. 1984, *ApJS*, **54**, 443.

Prialnik, D., & Kovetz, A. 1995, *ApJ*, **445**, 789.

Sarna, M.J., Marks, P.B., & Smith, R.C. 1995, *MNRAS*, **276**, 1336.

Schreiber, M.R., & Gänsicke, B.T. 2003, *A&A*, **406**, 305.

Schreiber, M.R., Gänsicke, B.T., & Cannizzo, J.K. 2000, *A&A*, **362**, 268.

Shara, M.M. 1989, *PASP*, **101**, 5.

Shara, M. M., Livio, M., Moffat, A. F. J., & Orio, M. 1986, *ApJ*, **311**, 163.

Warner, B. 1995, Cataclysmic Variable Stars (Cambridge: Cambridge Univ. Press).

Wilson, R.E. 1979, *ApJ*, **234**, 1054.

Wilson, R.E. 1990, *ApJ*, **356**, 613.

Wood, M.A. 1995, in White Dwarfs, ed. D. Koester,& K. Werner (New York: Springer), 41.

EVIDENCE FOR LARGE SUPERHUMPS IN TX COL AND V4742 SGR

Alon Retter
School of Physics
University of Sydney, 2006
Australia
retter@Physics.usyd.edu.au

Alexander Liu
Norcape Observatory
PO Box 300
Exmouth, 6707
Australia
asliu@onaustralia.com.au

Marc Bos
Mt Molehill Observatory
83a Hutton Street
Otahuhu, Auckland
New Zealand
cushla@kiwilink.co.nz

Abstract Since the discovery of the largest positive superhump period in TV Col (6.4 h), we have started a program to search for superhumps in cataclysmic variables (CVs) with large orbital periods. In this work, we summarize preliminary results of our observations of TX Col and V4742 Sgr. TX Col is an intermediate polar with a 5.7-h orbital period. V4742 Sgr is a recent (2002) nova with no known periods. CCD unfiltered continuous photometry of these two objects was carried out during 56 nights (350 hours) in 2002-2003. The time series analysis reveals the presence of several periods in both power spectra. In TX Col, in addition to the orbital period of 5.7 h, we found peaks at 7.1 h and 5.0 h. These are interpreted as positive and negative superhumps correspondingly, although the effects of the quasi-periodic oscillations at $\sim$2 h (which may cause spurious signals) were not taken into consideration. In the light curve of V4742 Sgr two long periods are detected – 6.1 and 5.4 h as well as a short-term period at 1.6 h. This result suggests that V4742 Sgr is an intermediate polar candidate and a permanent superhump system with a large orbital period (5.4 h) and a superhump

E.M. Sion, S. Vennes and H.L. Shipman (eds.), White Dwarfs: Cosmological and Galactic Probes, 251-259.
© 2005 *Springer*.

period excess of 13%. If these results are confirmed, TX Col and V4742 Sgr join TV Col to form a group of intermediate polars with extremely large superhump periods. There seems to be now growing evidence that superhumps can occur in intermediate polars with long orbital periods, which is very likely inconsistent with the theoretical prediction that superhumps can only occur in systems with mass ratios below 0.33. Alternatively, if the mass ratio in these systems is nevertheless below the theoretical limit, they should harbour undermassive secondaries and very massive white dwarfs, near the Chandrasekhar limit, which would make them excellent candidates for progenitors of supernovae type Ia.

Keywords: accretion disc, novae

1. Introduction

Binary systems often show quasi-periodicities a few percent longer than their orbital periods. They are understood as the beat periods between the orbital period and the apsidal precession of the accretion disc. For historical reasons they are known as positive superhumps. The positive superhumps obey a nice relation between the superhump period excess over the orbital period and the orbital period (Stolz & Schoembs 1984; Patterson 1999). Negative superhumps, quasi-periodicities a few percent shorter than the orbital periods, are explained by the beat periods between the orbital period and the nodal precession of the accretion disc. The negative superhumps follow a somewhat similar relation between the superhump period deficit over the orbital period and the orbital period (Patterson 1999, see also Retter 2002). Superhumps are important as the binary mass ratio can be estimated from the observed difference between the superhump and orbital period.

According to theory (Whitehurst & King 1991; Murray 2000) precessing accretion discs can occur only in binaries with small mass ratios (q=$M_2/M_1 \leq$ 1/3). In CVs, systems with longer orbital periods have larger separations and their secondaries are thus more massive since they have to fill their Roche Lobes. There is a small scatter on the mass of the primary white dwarf and therefore, the limit on the mass ratio is translated into orbital periods shorter than about 3-4 h. TV Col, with an orbital period of 5.5 h and a negative superhump of 5.2 h has been an unusual case. Retter et al. (2003) found another period, 6.4 h, in existing data of this object and confirmed it by further observations. It is naturally understood as a positive superhump. These results raised the question whether TV Col is unique. This work shows that it is almost certainly not.

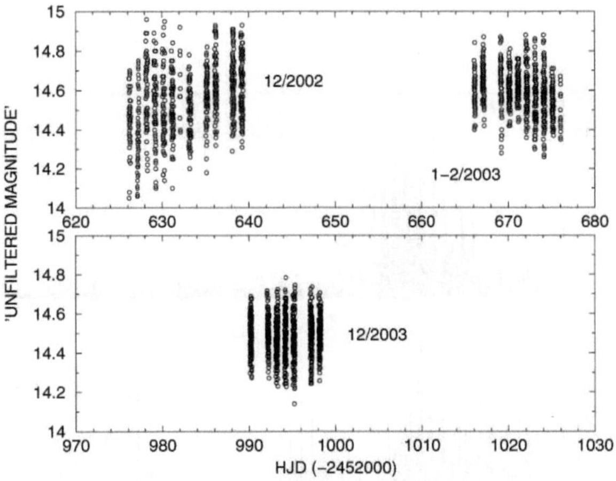

Figure 1. The light curve of TX Col during 22 nights in 2002-2003 and 7 nights in 2003.

2. Observations, Analysis and Results

TX Col

Photometric unfiltered CCD observations of TX Col, a 5.7-h orbital period CV (Buckley & Tuohy 1989; Norton et al. 1997), were carried out using a 0.3-m Meade LX200 telescope and a CCD in Norcape Observatory, Exmouth, Australia during 22 nights from December 2002 until February 2003 and in 7 nights in December 2003. No filter was used. Fig. 1 shows the light curve of this system.

Fig. 2 presents the power spectra (Scargle 1982) of these observations. Note that for the first observing season (December 2003 – February 2002) runs shorter than 5 h were rejected and the mean was subtracted from each night. In the power spectrum of the second season (December 2003) no de-trending method was use. The peaks at the right hand-side of the diagram (which are more evident in the observations of the second season) correspond to the spin period (1911 s, 45.2 d^{-1}) and its beat with the orbital period (2106 s, 41.0 d^{-1}). The light curve also has quasi-periodic oscillations at $\sim$2 h ($\sim$12 d^{-1}).

In Fig. 3 we show the frequency interval 1-7 d^{-1}. The data from the first season (top panel) show two groups of peaks in addition to the orbital period (f_3). The peak-to-peak amplitudes of these signals are about 0.1 mag. We propose that the 5.2-h period (f_1) is a negative superhump and that the 7.0-h period (f_2) – a positive superhump. 'a_i' (i=1-3) represent 1-d^{-1} aliases of 'f_i' correspondingly. The data from the second observing season confirms the presence of the 7.0-h period (f_2).

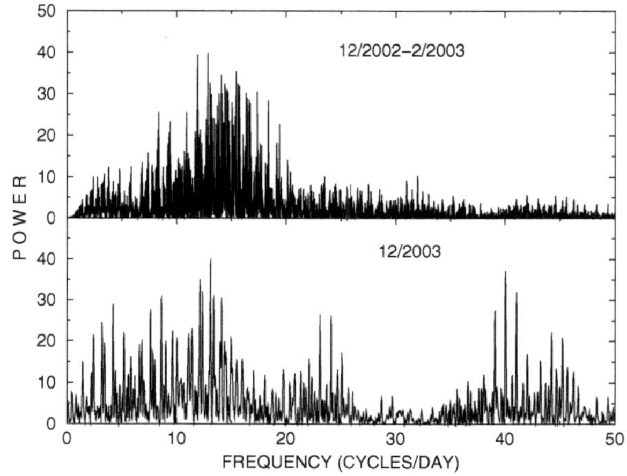

Figure 2. Power spectra of two observing seasons of TX Col in 2002-2003. Top panel: December 2002 – February 2003. Bottom panel: December 2003. See text for more details.

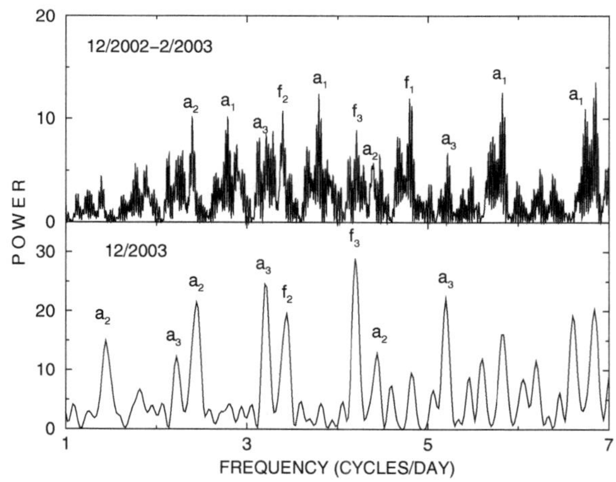

Figure 3. Same as Fig. 2 for the frequency interval 1-7 d^{-1}. f_3 is the known orbital period (5.7 h). In the power spectrum of the first observing run there are two additional groups of peaks centered around f_1 (5.2 h) and f_2 (7.0 h) which we interpret as negative and positive superhumps respectively. f_2 also appears in the data of the second season. See text for more details.

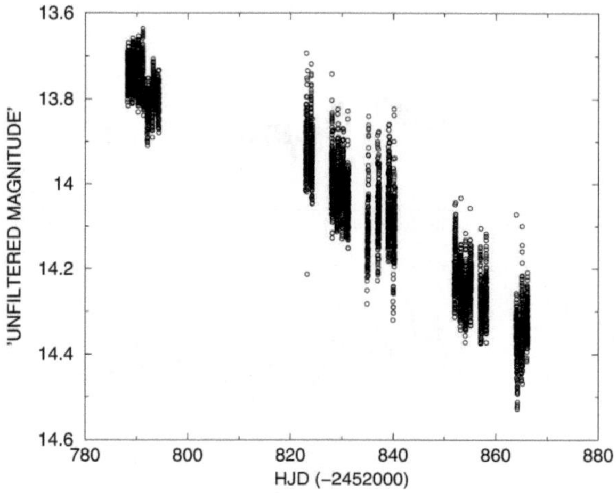

Figure 4. The unfiltered light curve of Nova V4742 Sgr 2002/2 during 26 nights in May-July 2003.

V4742 Sgr

Fig. 4 displays the photometric unfiltered CCD observations of Nova V4742 Sgr 2002/2 taken during 26 nights in May-July 2003. The nova decayed by about 0.6 mag during the observing run. The nightly light curves show erratic behaviour, which is typical of data which are modulated with several periods and which is similar to the light curves of TV Col and TX Col. Two long runs are shown as examples in Fig. 5.

The power spectrum of the whole data (after subtracting the long-term trend and the mean from each night) showed several peaks around 6 h, however it was somewhat noisy due to the presence of short runs. Therefore, runs shorter than 0.29 d ($\sim$7 h) were rejected. Figs. 6 and 7 present the power spectrum of the remaining 15 nights. Similar to the data of TX Col, the power spectrum of V4742 Sgr displays a complicated multi-periodic structure. We could identify two long-term frequencies in the data – 3.96 and 4.48 d^{-1}, which correspond to 6.1 and 5.4 h respectively. The peak-to-peak amplitudes of these signals are about 0.05 mag. The difference between the two periods ($\sim$13%) would fit a positive superhump excess if we interpret the 5.4 h peak as the orbital period and the 6.1 h peak as a positive superhump.

The presence of the 5.4 h period was confirmed by CCD unfiltered observations using a 0.25-m telescope in Mt Molehill Observatory, Auckland, New Zealand during 5 nights (25 h) in July-August 2003.

The power spectrum of V4742 Sgr also shows several peaks at shorter periods. This structure suggests that the nova is an intermediate polar system

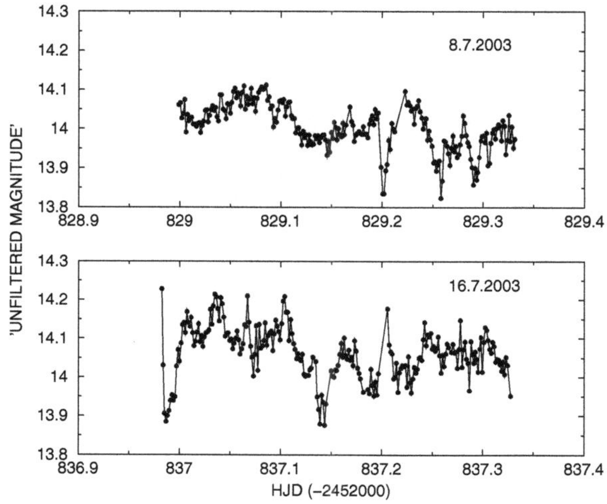

Figure 5. Two sample light curves of V4742 Sgr. Top panel: July 8th, 2003. Bottom panel: July 16th, 2003

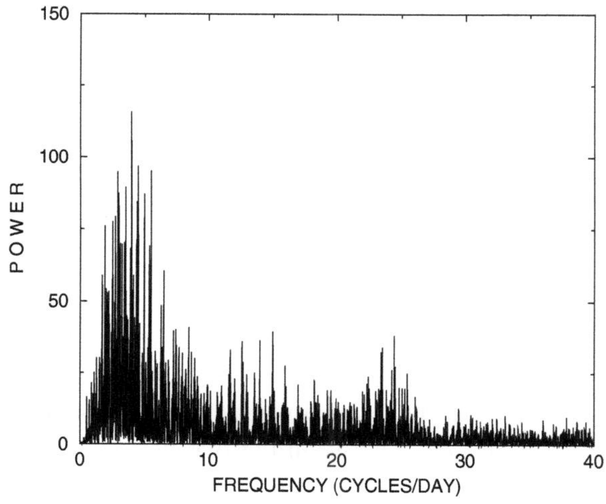

Figure 6. Power spectrum of the 15 longest runs of Nova V4742 Sgr 2002/2. It shows several peaks. See also Fig. 7.

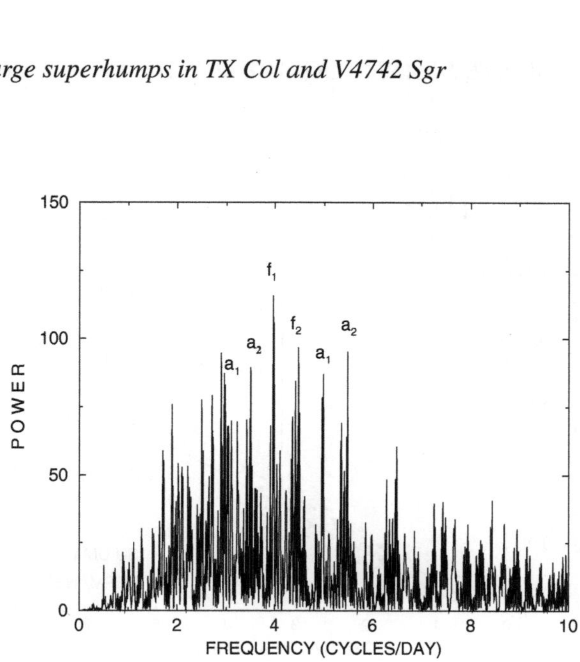

Figure 7. A zoom of Fig. 6 into the frequency interval 0-10 d^{-1}. The power spectrum shows two long-term periods and their 1-d^{-1} aliases. We suggest that the 5.4-h peak (f_2) is the orbital period and that the 6.1-h peak (f_1) is a positive superhump. a_i (i=1,2) correspond to 1-d^{-1} aliases of f_i respectively. In addition, the peaks at longer frequencies (Fig. 6) support an intermediate polar model for this system.

with a spin period of 1.6 h or 59 min. Confirmation by X-ray observations is naturally required for this suggestion.

Tests

The presence of the periodicities in the data of TX Col and V4742 Sgr was checked by several tests. We calculated the power spectrum of the air-masses; we subtracted one period and checked whether the other/s disappear from the power spectrum of the residuals; we planted the period/s in the data and checked its/their aliases. Simulations were also carried out to check the significance of one period in the presence of the other/s. The data were divided into subsections and the power spectra of the runs were compared. We also checked different de-trending methods (and in particular subtracting the trend from each night, which is especially relevant for the decaying nova, V4742 Sgr).

These tests support the above findings. We note, however, that for TX Col we did not try to estimate the influence of the quasi-periodic oscillations. Quantifying and simulating this effect is extremely hard. The presence of quasi-periodic oscillations may cause spurious signals. In the case of V4742 Sgr, this effect (if exists at all) is weak and cannot form the observed strong signals unless the periods themselves represent quasi-periodic oscillations.

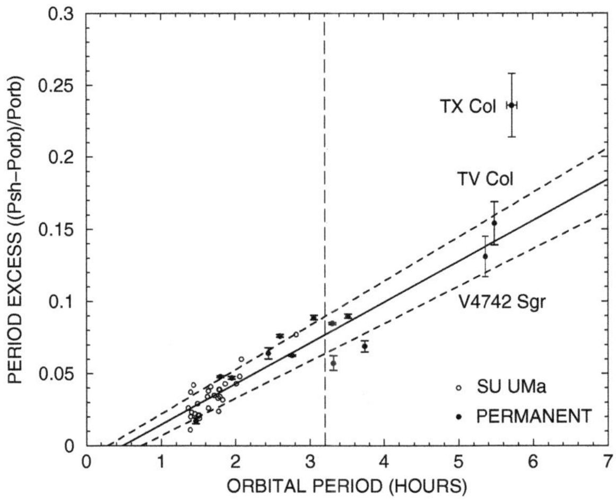

Figure 8. The relation between the positive superhump excess and the orbital period. TV Col and V4742 Sgr obey the relation while TX Col somewhat deviates from it. Note that the values for V4742 Sgr and TX Col still require confirmation.

Another warning comment is that the detected signals are near the length of the nightly runs. We tried to overcome this problem by rejecting short runs, however, for TX Col observations shorter than 5 h (shorter than the suggested periods) were still included. For V4742 Sgr, nights shorter than 7 h were rejected, therefore the remaining nights are longer than the proposed periods. This means that our results of V4742 Sgr stand on a safer ground. Anyway, it is recommended to confirm these results by further, multi-site observations.

3. Discussion

The analysis of the data of TX Col and V4742 Sgr reveals strong evidence that they have several periods in their light curves. The uncertainties in these findings were outlined in the previous section. Assuming that these results are real, and adopting our interpretation of the periods of TX Col and V4742 Sgr, we plotted in Fig. 8 the extension of the relation for positive superhumps to long periods. TV Col and V4742 Sgr obey the relation while TX Col deviates from it having a superhump period excess somewhat larger than the predicted value. This diagram suggests that superhumps may be common in CVs with orbital periods up to about 6 h (or even larger).

4. Summary

This work gives evidence that TX Col and V4742 Sgr have several large periods and thus supports the idea that superhumps can be found in CVs with large orbital periods. We note, however, that the light curve of TX Col shows quasi-periodic oscillations with large amplitudes that complicate the analysis and may cause spurious signals. The data of V4742 Sgr do not show this behaviour, however, to firmly state that the nova is a permanent superhump system with a large orbital period its orbital period should be confirmed by a radial velocity study. The analysis of the photometric observations of V4742 Sgr also indicate that it may be an intermediate polar system. Thus, we feel that there is growing evidence that superhumps can occur in CVs with orbital periods of 5-6 h. This almost certainly means that they have mass ratios larger than the theoretical limit of 0.33. The reason for this behaviour may be that they all all intermediate polars. The presence of superhumps with large periods may be alternatively understood if the systems have undermassive secondary stars and massive primary white dwarf. In this case, these systems are excellent candidates for supernovae type IA.

We encourage observers to look for superhumps is CVs with large orbital periods including dwarf novae above the period gap.

Acknowledgments

AR is supported by a grant from the Australian Research Council.

References

Buckley, D.A.H., and Tuohy, I.R. 1989, *ApJ*, **344**, 376.

Murray, J.R. 2000, *MNRAS*, **314**, L1.

Norton, A.J., Hellier, C., Beardmore, A.P., Wheatley, P.J., Osborne, J.P., and Taylor, P. 1997, *MNRAS*, **289**, 362.

Patterson, J. 1999, in Disk Instabilities in Close Binary Systems, eds. S. Mineshige, C. Wheeler (Universal Academy Press, Tokyo), 61.

Retter, A., Chou, Y., Bedding, T., and Naylor, T. 2002, *MNRAS*, **330**, L37.

Retter, A., Hellier, C., Augusteijn, T., Naylor, T., Bedding, T., Bembrick, C., McCormick, J., and Velthuis, F. 2003, *MNRAS*, **340**, 679.

Scargle, J.D. 1982, *ApJ*, **263**, 835.

Stolz, B., and Schoembs, R. 1984, *A&A*, **132**, 187.

Whitehurst, R., and & King, A. 1991, *MNRAS*, **249**, 25.

CONCLUDING REMARKS

Understanding the Ultimate Fate of the Universe

Harry Shipman
Department of Physics and Astronomy, University of Delaware, Newark, DE19710, USA

Edward Sion
Department of Astronomy and Astrophysics, Villanova University
800 Lancaster Ave, Villanova, PA 19085, USA

Stéphane Vennes
Department of Physics and Astronomy, Johns Hopkins University
Baltimore, MD 21218-2686, USA

Some readers turn to the last chapter in a book in order to get an overview of the big message that the book contains. Those readers include a few intrepid souls who began the book at page 1 and have finally come to this point. Other readers, like ourselves, often begin an encounter with some kinds of written material by looking at the end in order to see the big picture. This brief final summary is for such readers.

Our purpose in holding the conference and assembling the book was to:

- inform astronomers, graduate students, and anyone else who encountered this material about some of the exciting research being done by scientists studying white dwarf stars and cataclysmic variable stars, the late stages of the life cycles of low-mass stars like our sun; and
- elicit some connections between studies of late stages of stellar evolution and broader questions like the age of the Milky Way Galaxy and the use of type Ia supernovae as standard candles to determine the ultimate fate of the universe.

We will resist the tendency of conference summarizers to list the papers which struck us as being particularly interesting or worthy of note. Our list, like anyone's list, is idiosyncratic. We will, however, quote a few words from our invited reviewers which best summarize the state of affairs. Briefly, we will simply make the general observation that our understanding of the cooling

E.M. Sion, S. Vennes and H.L. Shipman (eds.), White Dwarfs: Cosmological and Galactic Probes, 261-263.
© 2005 *Springer*.

of white dwarf stars and the evolution of white dwarf stars in close binaries has moved forward considerably in the past several years but still has a way to go if it is to become a firm foundation for other areas of astronomical work. Our understanding of the basic physics of white dwarf cooling is on firmer ground than it was a decade or so ago. However, there are still a number of gaps in our knowledge, both observational and theoretical: Where does the white dwarf sequence really stop? How does this termination differ among the thick disk, thin disk, and halo populations? How do some of the unknown factors in the interior structure of white dwarf stars affect our determination of cooling ages for a particular white dwarf population? What better way to summarize the situation than quoting Fontaine, Bergeron, and Brassard concerning the detection of white dwarfs old enough to be of cosmological interest:

> Except for the well documented turnover in the luminosity function of local white dwarfs at low temperatures which is directly related to the finite age of the galactic disk, the optimistic view that many shared several years ago has subdued somewhat as these populations have remained elusive.

The use of type Ia supernovae to determine that the universe is accelerating and contains dark energy has been a major focus of research for the past ten years and has been described in many reviews. But our understanding falls short in a very basic way: We have a generalized picture of their origin namely the collapse of a white dwarf star in a close binary system. And while that picture is consistent with what we know, we have no direct evidence that can connect that generalized picture with real stellar systems. Quoting Filippenko:

> A number of possible systematic effects (dust, supernova evolution) thus far do not seem to eliminate the need for $\Omega_\Lambda > 0$. However, during the past few years some very peculiar low-redshift SNe Ia have been discovered, and we must be mindful of possible systematic effects if such objects are more abundant at high redshifts.

Despite much effort, we have yet to identify a binary system that will become a Type Ia supernova on a reasonably short time scale. We would like to make a convincing argument that a particular class of CVs are sufficiently numerous that they can be reliably identified as the progenitors of Type Ia supernovae. Unfortunately, we are not there yet, and quoting Tout:

> It may well be that the true nature of the progenitors has not yet even been conceived of.

The long term accretion in cataclysmic variables with massive white dwarfs may yet prove to be an effective channel for supernova events. Advances in understanding the effect of accretion during long term evolution have been achieved by comparing measured CV white dwarf effective temperatures during quiescence/low brightness states with time-dependent models of boundary layer irradiation and compressional heating of the white dwarfs. However, the mean value of $\dot{M}$ between nova eruptions remains uncertain. There is strong evidence that hibernation on time scales of $\sim 10^3$ years does occur but to quote Brian Warner and Patrick Woudt concerning accretion in cataclysmic variables:

> Until indirect means of estimating the duty cycle of this (hibernation) process are found, the value of $\dot{M}$ will remain very uncertain.

The rotational velocities or upper limits of 15 accreting white dwarfs are known from *HST* STIS and *FUSE* observations, thus providing critical insight into angular momentum transfer during long term disk accretion. While the CV white dwarfs are very rapidly rotating relative to single white dwarfs, so far all are rotating considerably smaller than breakup. Perhaps the accreted angular momentum is lost every time a nova shell is ejected. To quote Szkody, Sion and Gaensicke:

> It is apparent that spin-up from accretion is counterbalanced by spin losses through other (as yet unidentified) mechanisms.

Perhaps the most encouraging aspect of the conference was a strong reminder of the importance of this kind of research for its own sake. Suppose that cosmologists come up with a credible model of the universe that fits a reasonable enough number of datasets to several decimal places and achieves the status of an explanation that is so credible that it is not seriously questioned. It is, then, still essential to understand the way that white dwarfs cool and the way that close binaries with white dwarf component(s) make Type Ia supernovae. These are important stellar objects, precisely because they do give us insight into the nature of the Universe at very early times. We cannot understand the Universe at 0.2 Hubble times without understanding the nature of the objects that were around at that time. Research work in the stellar graveyard is still worth doing.

Object Index

Author Index

MIX
Papier aus verantwortungsvollen Quellen
Paper from responsible sources
FSC® C105338

If you have any concerns about our products,
you can contact us on
ProductSafety@springernature.com

In case Publisher is established outside the EU,
the EU authorized representative is:
Springer Nature Customer Service Center GmbH
Europaplatz 3, 69115 Heidelberg, Germany

Printed by Libri Plureos GmbH
in Hamburg, Germany